SpringerBriefs in Applied Sciences and Technology

Computational Intelligence

Series Editor

Janusz Kacprzyk, Systems Research Institute, Polish Academy of Sciences, Warsaw, Poland

SpringerBriefs in Computational Intelligence are a series of slim high-quality publications encompassing the entire spectrum of Computational Intelligence. Featuring compact volumes of 50 to 125 pages (approximately 20,000-45,000 words), Briefs are shorter than a conventional book but longer than a journal article. Thus Briefs serve as timely, concise tools for students, researchers, and professionals.

More information about this subseries at http://www.springer.com/series/10618

Patricia Melin · Juan Carlos Guzmán ·
German Prado-Arechiga

Neuro Fuzzy Hybrid Models for Classification in Medical Diagnosis

Patricia Melin
Division of Graduate Studies
Tijuana Institute of Technology
Tijuana, Baja California, Mexico

Juan Carlos Guzmán
Division of Graduate Studies
Tijuana Institute of Technology
Tijuana, Baja California, Mexico

German Prado-Arechiga
Department of Cardiodiagnostico
Excel Medical Center
Tijuana, Baja California, Mexico

ISSN 2191-530X   ISSN 2191-5318   (electronic)
SpringerBriefs in Applied Sciences and Technology
ISSN 2625-3704   ISSN 2625-3712   (electronic)
SpringerBriefs in Computational Intelligence
ISBN 978-3-030-60480-6   ISBN 978-3-030-60481-3   (eBook)
https://doi.org/10.1007/978-3-030-60481-3

This Springer imprint is published by the registered company Springer Nature Switzerland AG
The registered company address is: Gewerbestrasse 11, 6330 Cham, Switzerland

Preface

Nowadays, the use of artificial intelligence techniques, such as fuzzy logic, neural networks and evolutionary computing, help to design efficient models that can give an accurate diagnosis and, therefore, facilitate decision-making; based on experience and the rules established by experts in the area, this research focused on the diagnosis of blood pressure.

It is difficult to determine if we suffer from high blood pressure. For this reason, it is important to constantly have checkups; one of the most recommended checks is the 24 h monitoring in which a report of the blood pressure behavior is obtained during the next 24 h. This is why we worked to form a solid database with real monitoring to perform tests with the proposed model.

Once the database of the considered patients is collected, a neural network model analyzes this information. The neural network is responsible for modeling a 24 h monitoring data of a patient and based on this information provide a tendency for the patient's blood pressure. In particular, the systolic and diastolic inputs of the fuzzy classifier are obtained, which based on the knowledge of an expert and guidelines of blood pressure already specified by cardiologists will produce a fast and correct diagnosis.

Finally, the diagnosis is presented in a graphic interface in which the information is analyzed and interpreted in a friendly way. The use of the graphical interface helps the expert with the interpretation of the analyzed data, to graphically show the behavior of the data and a final diagnosis.

The main contribution of this book is to provide a new fuzzy neural hybrid model for classification in medical diagnosis. This model is able to analyze and interpret the data collected from a patient and provide a correct diagnosis, which can help avoid complications in the long term.

This book is intended to be a reference for scientists and engineers interested in applying intelligent techniques for solving problems in medicine. We believe that this book can serve as a reference for further research or to continue with what is proposed in this book for improving the results.

In Chap. 1, we begin by offering an introduction of the use of intelligent computing techniques in medicine, which is becoming more common, and some of them are neural networks, fuzzy logic and evolutionary computation. The design and implementation of the proposed classifiers are described. We also mentioned that some research works have been done to diagnose blood pressure using intelligent techniques.

We describe in Chap. 2 some basic concepts of blood pressure, which are very important; these concepts help to understand a more about this work. Also, Computational Intelligence Techniques such as fuzzy logic, neural networks and evolutionary computing are mentioned.

We describe in Chap. 3 the proposed method that is based on the neural fuzzy model for classification in medical diagnosis. In this book, it is being applied to blood pressure for the classification of blood pressure, where it is determined whether a patient suffers from hypertension or not.

Chapter 4 describes different cases studies that are analyzed in this book for the development of the proposed model. The design of fuzzy classifiers and optimization of membership functions and rules is also presented. In addition, comparative studies between different guidelines of blood pressure classification are presented. The statistical test is also shown for each case study.

We explained in Chap. 5 the conclusions of this work are that we have developed a new model using neuro-fuzzy hybrid techniques that actually implements the human reasoning using a set of decision rules for the study of different diseases such as hypertension blood pressure (HBP). This new neuro-fuzzy hybrid model (NFHM) provides us a faster, safer and accurate tool for an objective diagnostic without inter-observer variability, in this case based on the classification of hypertension, according to the definitions of the European Guidelines. This method is very efficient, therefore takes less time and is more accurate for classify the level of HBP.

In general, the performance of the neuro-fuzzy hybrid model so far has a good functioning, but it is necessary to continue with more experiments and have complete databases to be able to help the model to approximate and classify the information in the best possible way. As future work, we would increase the number of patients in the current database. Apply the proposed model in other cardiovascular diseases to support the final diagnosis.

We end this preface of the book by giving thanks to all the people who have helped in this project and during the writing of this book. We would like thank our funding institutions CONACYT and TNM of our country for the support within this project. We have to thank our institution, Tijuana Institute of Technology, for always supporting our projects. Finally, we thank our families for the support and patience in this project.

Tijuana, Mexico   Prof. Patricia Melin
                                                                    Dr. Juan Carlos Guzmán
                                                                Dr. German Prado-Arechiga

Contents

Chapter 1
Introduction to Neuro Fuzzy Hybrid Model

Today, the use of intelligent techniques in medicine is becoming more common, and just to mention a few: neural networks, fuzzy logic and evolutionary computation [1–6]. The main idea of the book is to describe a fuzzy neural hybrid model that can make a diagnosis, which can be fast and accurate and, therefore, it is necessary to have a 24-h patient monitoring database. The main contribution of this work is the contribution of the neuro fuzzy hybrid model shown in Fig. 3.1, which consists of an information input module (database), then a neural network that models the data obtained and finally provides a trend based on the data learned and modeled. Then the design of a fuzzy classifier was designed based on the knowledge of an expert, to later be optimized by some algorithm to improve the membership functions and rules provided by the expert. Some important works done for this neuro fuzzy model can be seen in the following papers: Design of an optimized fuzzy classifier for the diagnosis of blood pressure with a new computational method for expert rule optimization and Optimal Genetic Design of Type-1 and Interval Type-2 Fuzzy Systems for Blood Pressure Level Classification, without belittling the other work done [7–14]. Another goal is to find a modular neural network architecture, which will help to have an accurate model to know the blood pressure trend of the patient, and a fuzzy classifier is also needed, which will classify the blood pressure level to which the patient belongs [15–17]. First, it is important to focus on the design of the fuzzy classifier to classify the blood pressure level. The outputs of the modular neural network are the systolic pressure and the diastolic pressure, which will be used as inputs for the fuzzy classifier. The results of the proposed classifier are expected to be faster and more accurate [18–24]. Currently there are few works done based on artificial intelligence techniques for the diagnosis of blood pressure and regularly in most of these works, the classification is generally, for example: low, medium and high. In this proposed work we used hypotension, optimal, normal, high normal, grade 1, 2, 3 hypertension and isolated grade 1, 2, 3 hypertensions as indicated by the guidelines

P. Melin et al., *Neuro Fuzzy Hybrid Models for Classification in Medical Diagnosis*, SpringerBriefs in Computational Intelligence, https://doi.org/10.1007/978-3-030-60481-3_1

1

of the European Society of Cardiology shown in Table 2.1. Some research work to diagnose blood pressure has been done using intelligent techniques, for example the paper of Hypertension Diagnosis using a Fuzzy Expert System, in this article a fuzzy expert system was designed to assess the risk of hypertension in a patient using risk factors and blood pressure, the expert system results in the risk and not the blood pressure level [25]. In this paper, a Genetic Neuro Fuzzy System for Hypertension Diagnosis use a backpropagation network and a genetic algorithm to initialize the neuro fuzzy system, and the systems only diagnose the risk of hypertension and not the classification levels [26]. The next work is a neural network expert system for diagnosing and treating hypertension, in this paper, a Model of Neural Network was designed for diagnosis and treated of hypertension and also for building models of "Hypernet" using as expert system [27]. These works differ in that they only use the intelligent techniques to evaluate the hypertension risk diagnosis. In this book, we construct a complete model for classification of medical diagnosis focused on the trend of the blood pressure of patient given by the neural network and the blood pressure load using readings given by 24-h monitoring [28–32].

The Medical Doctor aims at achieving the best possible diagnosis for the patient, therefore, the implementation of an appropriate method that can model this problem allows the doctor to provide the best possible diagnosis. The diagnosis of hypertension is a very important topic in medicine since hypertension is a disease that in some occasions is silent and is therefore very dangerous, thus threatening human health. It is a disease that usually has fatal results, such as stroke, heart attack and kidney failure [33–42].

One of the main dangerous aspects of hypertension is that it is a silent disease and people have no idea that they have it, a third of people with high blood pressure do not know they have hypertension. The only way to know if your blood pressure is high is through regular checkups.

In Medicine it is very common to use models applied to the diagnosis of future diseases and thus be able to treat them in an appropriate way. The implementation of an appropriate method in Medicine can be a very important point since it helps to treat diseases in time and save lives [43–45]. The data obtained as a future trend is modeled to help decision-making in the medium and long term, due to the precision or inaccuracy of the modeled data allows for patient control [46–51].

It is important to know the future behavior of a patient's blood pressure, this allows for better decisions to be made to better diagnose of the patient's health and avoid future problems, which lead to premature death due to not having adequate treatment [52–57].

This book is organized as follows: In Chap. 2 the theory and background is shown, in Chap. 3 the proposed Neuro fuzzy hybrid model is presented, Chap. 4 shows the study cases to test the neuro fuzzy hybrid model studies, and Chap. 5 offers the conclusions of the neuro fuzzy hybrid model and future work about the work presented in this book.

References

1. Yang, X. S., Karamanoglu, M., & He, X. (2014). Flower pollination algorithm: A novel approach for multiobjective optimization. *Engineering Optimization, 46,* 1222–1237.
2. Yu, J. J. Q., & Li, V. O. K. (2015). A social spider algorithm for global optimization. *Applied Soft Computing, 30,* 614–627.
3. Meng, X.-B., Gao, X. Z., Lu, L., Liu, Y., & Zhang, H. (2016). A new bio-inspired optimisation algorithm: Bird swarm algorithm. *Journal of Experimental and Theoretical Artificial Intelligence, 28,* 673–687.
4. Gopinathannair, R., & Olshansky, B. (2015). Management of tachycardia. *F1000Prime Reports, 7,* 60.
5. Wilson, J. M. (2005). Essential cardiology: Principles and practice. *Texas Heart Institute Journal, 32,* 616.
6. Lai, C., Coulter, S.A., & Woodruff, A. (2017). Hypertension and pregnancy. *Texas Heart Institute Journal, 44*(5), 350–351.
7. Guzman, J. C., Melin, P., & Prado-Arechiga, G. (2017). Design of an optimized fuzzy classifier for the diagnosis of blood pressure with a new computational method for expert rule optimization. *Algorithms, 10,* 79.
8. Guzmán, J. C., Melin, P., & Prado-Arechiga, G. (2017). Neuro-fuzzy hybrid model for the diagnosis of blood pressure. In P. Melin, O. Castillo, & J. Kacprzyk (Eds.), *Nature-inspired design of hybrid intelligent systems* (pp. 573–582). Cham, Switaerland: Springer International Publishing.
9. Guzmán, J.C., Melin, P., & Prado-Arechiga, G. (2015). Design of a fuzzy system for diagnosis of hypertension. In *Design of intelligent systems based on fuzzy logic, neural networks and nature-inspired optimization*, pp. 517–526. Cham, Switaerland: Springer International Publishing.
10. Guzmán, J.C., Melin, P., & Prado-Arechiga, G. (2016). Artificial intelligence utilizing neuro-fuzzy hybrid model for the classification of blood. European Society of Hypertension. *Journal of Hypertension, 34.*
11. Guzmán, J.C., Melin, P., & Prado-Arechiga, G. (2016). Classification of blood pressure based on a neuro-fuzzy hybrid computational model. European Society of Hypertension. *Journal of Hypertension, 34.*
12. Guzmán, J.C., Melin, P., & Prado-Arechiga, G. (2017). An interval type-2 fuzzy logic approach for diagnosis of blood pressure. *Journal of Hypertension, 34.*
13. Guzmán, J. C., Melin, P., & Prado-Arechiga, G. (2017). Design of an optimized fuzzy classifier for the diagnosis of blood pressure with a new computational method for expert rule optimization. *Algorithms, 10*(3), 79. https://doi.org/10.3390/a10030079.
14. Guzmán, J.C., Melin, P., Prado-Arechiga, G., & Miramontes, I. (2018). A comparative study between european guidelines and American guidelines using fuzzy systems for the classification of blood pressure. *Journal of Hypertension, 36.*
15. Zadeh, L. A. (1965). Fuzzy sets. *Information and Control, 8,* 338–353.
16. Carvajal, O. R., Castillo, O., & Soria, J. J. (2018). Optimization of membership function parameters for fuzzy controllers of an autonomous mobile robot using the flower pollination algorithm. *Journal of Automation, Mobile Robotics and Intelligent Systems, 12,* 44–49.
17. Guzmán, J. C., Miramontes, I., Melin, P., & Prado-Arechiga, G. (2019). Optimal genetic design of type-1 and interval type-2 fuzzy systems for blood pressure level classification. *Axioms, 8,* 8.
18. Guzmán, J.C., Melin, P., & Prado-Arechiga, G. (2015). Design of a fuzzy system for diagnosis of hypertension. In *Design of intelligent systems based on fuzzy logic, neural networks and nature-inspired optimization* (pp. 517–526). Springer International Publishing.
19. Guzmán, J.C., Melin, P., & Prado-Arechiga, G. (2016). A proposal of a fuzzy system for hypertension diagnosis. In *Novel developments in uncertainty representation and processing* (pp. 341–350). Springer International Publishing.

20. American Heart Association (2015). Available Online http://www.heart.org/HEARTORG/ Conditions/HighBloodPressure/High-Blood-Pressure-or-Hypertension_UCM_002020_Sub HomePage.jsp. Accessed on 9 July 2016.
21. Kenney, L., Humphrey, R., Mahler, D., & Brayant, C. (1995). *ACSM's guidelines for exercise testing and prescription.* Philadelphia, PA, USA: Williams & Wilkins.
22. Mangrum, J. M., & DiMarco, J. P. (2000). The evaluation and management of bradycardia. *New England Journal of Medicine, 342,* 703–709.
23. Mancia, G., Grassi, G., & Kjeldsen, S. E. (2008). *Manual of hypertension of the european society of hypertension.* London, UK: Informa Healtcare.
24. Wizner, B., Gryglewska, B., Gasowski, J., Kocemba, J., & Grodzicki, T. (2003). Normal blood pressure values as perceived by normotensive and hypertensive subjects. *Journal of Human Hypertension, 17,* 87–91.
25. Kaur, R., & Kaur, A. (2014). Hypertension diagnosis using fuzzy expert system. In *International Journal of Engineering Research and Applications (IJERA) National Conference on Advances in Engineering and Technology, AET*, 29th March 2014.
26. Kaur, A., Bhardwaj, A., & Been, U.A.H. (2014). Genetic neuro fuzzy system for hypertension diagnosis. *Heart, 19,* 25.
27. Poli, R., et al. (1991). A neural network expert system for diagnosing and treating hypertension. *Computer, 24*(3), 64–71.
28. Sikchi, S., & Ali, M. (2013). Design of fuzzy expert system for diagnosis of cardiac diseases. *International Journal of Medical Science and Public Health, 2,* 56.
29. Rosendorff, C. (2013). *Essential cardiology* (3rd ed.). Bronx, NY, USA: Springer.
30. Melin, P., & Castillo, O. (2005). *Hybrid intelligent systems for pattern recognition using soft computing.* Berlin/Heidelberg, Germany: Springer-Verlag.
31. Asl, A.A.S., & Zarandi, M.H.F. (2017). A type-2 fuzzy expert system for diagnosis of Leukemia. In *Fuzzy logic in intelligent system design, proceedings of the North American fuzzy information processing society annual conference*, Cancun, Mexico, 16–18 October 2017. Springer, Cham, Switzerland, 2017, pp. 52–60.
32. Sotudian, S., Zarandi, M.H.F., & Turksen, I.B. (2016). From type-I to type-Ii fuzzy system modeling for diagnosis of hepatitis. *World Academy of Science, Engineering and Technology International Journal of Computer Electrical Automation Control and Information Engineering, 10,* 1280–1288.
33. Miramontes, I., Martínez, G., Melin, P., & Prado-Arechiga, G. (2017). A hybrid intelligent system model for hypertension risk diagnosis. In *Fuzzy logic in intelligent system design, proceedings of the North American fuzzy information processing society annual conference*, Cancun, Mexico, 16–18 October 2017. Springer, Cham, Switzerland, 2017, pp. 202–213.
34. Melin, P., Miramontes, I., & Prado-Arechiga, G. (2018). A hybrid model based on modular neural networks and fuzzy systems for classification of blood pressure and hypertension risk diagnosis. *Expert Systems with Applications, 107,* 146–164.
35. Miramontes, I., Martínez, G., Melin, P., & Prado-Arechiga, G. (2017). A hybrid intelligent system model for hypertension diagnosis BT. In P. Melin, O. Castillo, & J. Kacprzyk (Eds.), *Nature-Inspired design of hybrid intelligent systems* (pp. 541–550). Cham, Switaerland: Springer International Publishing.
36. Zarandi, M. H. F., Khadangi, A., Karimi, F., & Turksen, I. B. (2016). A computer-aided type-II fuzzy image processing for diagnosis of meniscus tear. *Journal of Digital Imaging, 29,* 677–695.
37. Pabbi, V. (2015). Fuzzy expert system for medical diagnosis. *International Journal of Science and Results Publication, 5,* 1–7.
38. Mohamed, K. A., & Hussein, E. M. (2016). Malaria parasite diagnosis using fuzzy logic. *International Journal of Scientific Research, 5,* 2015–2017.
39. Melin, P., & Prado-Arechiga, G. (2018). *New hybrid intelligent systems for diagnosis and risk evaluation of arterial hypertension.* Cham, Switzerland: Springer.
40. O'Brien, E., Parati, G., & Stergiou, G. (2013). Ambulatory blood pressure measurement. *Hypertension, 62,* 988–994.

41. Słowiński, K. (1992). Rough classification of HSV patients. In Intelligent decision support. In R. Słowiński (Ed.), *Theory and decision Library (Series D: System Theory, Knowledge Engineering and Problem Solving)* (Vol. 11). Dordrecht, The Netherlands: Springer.
42. Yuksel, S., Dizman, T., Yildizdan, G., & Sert, U. (2013). Application of soft sets to diagnose the prostate cancer risk. *Journal of Inequalities and Application, 2013,* 229.
43. Galilea, E. H., Santos-García, G., & Suárez-Bárcena, I. F. (2007). Identification of glaucoma stages with artificial neural networks using retinal nerve fibre layer analysis and visual field parameters. In E. Corchado, J.M. Corchado, & A. Abraham (Eds.), *Innovations in hybrid intelligent systems* (Vol. 44). *Advances in soft computing.* Berlin/Heidelberg, Germany: Springer.
44. Alcantud, J. C. R., Santos-García, G., & Hernández-Galilea, E. (2015). Glaucoma diagnosis: A soft set based decision making procedure. In J. Puerta (Ed.), *Advances in artificial intelligence, proceedings of the conference of the Spanish Association for Artificial Intelligence,* Vol. 9422. Albacete, Spain, 9–12 November 2015; Lecture Notes in Computer Science. Springer, Cham, Switzerland.
45. Alcantud, J. C., Biondo, Alessio E., & Giarlotta, A. (2018). Fuzzy politics I: The genesis of parties. *Fuzzy Sets and Systems, 349,* 71–98.
46. Texas Heart Institute. (2017). High blood pressure (hypertension) [Online]. Available https://www.texasheart.org/heart-health/heart-information-center/topics/high-blood-pressure-hypertension/. Accessed 08 October 2018.
47. Framingham Heart Study. (2019) [Online]. Available https://www.framinghamheartstudy.org/risk-functions/hypertension/index.php. Accessed 15 July 2019.
48. Bakris, G. L., & Sorrentino, M. (2017). *Hypertension: A companion to Braunwald's heart disease E-Book.* Elsevier Health Sciences.
49. Cain, G. (2017). *Artificial neural networks: New research.* New York: Nova Science Publishers, Incorporated.
50. Jin, L., Li, S., Yu, J., & He, J. (2018). Robot manipulator control using neural networks: A survey. *Neurocomputing, 285,* 23–34.
51. Saadat, J., Moallem, P., & Koofigar, H. (2017). Training echo estate neural network using harmony search algorithm. *International Journal of Artificial Intelligence, 15*(1), 163–179.
52. Villarrubia, G., De Paz, J. F., Chamoso, P., & De la Prieta, F. (2018). Artificial neural networks used in optimization problems. *Neurocomputing, 272,* 10–16.
53. Aggarwal, C. C. (2018). *Neural networks and deep learning: A textbook* (1st edn.), Cham: Springer International Publishing.
54. Shell, J., & Gregory, W. D. (2017). Efficient cancer detection using multiple neural networks. *IEEE Journal of Translation Engineering and Health and Medicine, 5,* 2800607.
55. Sadek et al., R.M. (2019). Parkinson's disease prediction using artificial neural network. *International Journal of Academy Health and Medicine Research, 3*(1), 1–8.
56. Pulido, M., Melin, P., & Mendoza, O. Optimization of ensemble neural networks with type-1 and interval type-2 fuzzy integration for forecasting the taiwan stock exchange.
57. Soto, J., Melin, P., & Castillo, O. (2018). A new approach for time series prediction using ensembles of IT2FNN models with optimization of fuzzy integrators. *International Journal of Fuzzy Systems, 20*(3), 701–728.

Chapter 2
Theory and Background of Medical Diagnosis

The following basic concepts are very important. These concepts help to understand more about this work.

2.1 Blood Pressure

The heart is a muscle that can pump blood to the tissues and organs of the body. Blood is pumped from the left side of the heart to the arteries, and the blood vessels carry oxygen and nutrients through the blood to the body. Blood pressure peaks when the heart muscle contracts and pumps blood is a tension called systolic. Diastolic stress is when the heart relaxes and fills with blood. Systolic stress is the top number and diastolic stress is the bottom number, for example 120/80 mmHg.

2.1.1 Type of Blood Pressure Diseases

It is the most common disease and currently increases both the morbidity and the mortality of cardiovascular diseases. There are different types of hypertension when the disease is subcategorized. These types are shown in Table 2.1 [1].

In Table 2.1, the blood pressure (BP) category is defined by the highest to lowest BP level, in this case systolic or diastolic blood pressure are managed. Isolated systolic hypertension is classified as 1, 2 or 3 according to the value of systolic BP in the indicated ranges.

© The Author(s), under exclusive license to Springer Nature Switzerland AG 2021
P. Melin et al., *Neuro Fuzzy Hybrid Models for Classification in Medical Diagnosis*,
SpringerBriefs in Computational Intelligence,
https://doi.org/10.1007/978-3-030-60481-3_2

Table 2.1 Definitions and classification of blood pressure levels

Category	Systolic		Diastolic
Hypotension	<90	And/or	<60
Optimal	<120	And	<80
Normal	120–129	And/or	80–84
High normal	130–139	And/or	85–89
Grade 1 hypertension	140–159	And/or	90–99
Grade 2 hypertension	160–179	And/or	100–109
Grade 3 hypertension	≥180	And/or	≥110
Isolated systolic hypertension	≥140	And	<90

2.1.2 Hypotension

It happens when the blood pressure is much lower than normal. This means that the heart, brain, and other parts of the body do not receive enough blood. Normal blood pressure is almost always between 90/60 and 120/80 mmHg. The medical name for low blood pressure is hypotension.

Blood pressure changes from person to person, a blood pressure measurement of less than 90 mm mercury (mmHg) of systolic blood pressure (highest pressure) or less than 60 mmHg of diastolic blood pressure (lowest number) is normally classified as low blood pressure.

Hypotension can have different causes, which can range from dehydration to serious medical or surgical disorders. This type of low blood pressure can be treated as long as the reason why the disease is happening is known [1].

2.1.3 Hypertension

This is a condition in which the pressure of the blood to the walls of the artery is too high.

Hypertension is generally defined as blood pressure above 140/90 and is considered severe when it is above 180/120.

In some cases, high blood pressure has no symptoms. If left untreated, over time, it can lead to health disorders, such as heart disease and stroke.

Following a healthy diet with less salt, exercising regularly, and taking medications can help lower blood pressure.

Cardiovascular disease is the leading cause of death worldwide today. However, the disease can be treatable but failure to comply with the recommendations given by the doctor can generate serious complications in the patient, such as bleeding or cerebral thrombosis, a myocardial infarction, which can be avoided by following the recommendations in the form correct.

The main consequence of hypertension is suffered by the arteries that are the ones that support the high blood pressure that is being generated continuously and this causes them to get thicker, the arteries become thicker and can damage the passage of blood through them and this is called arteriosclerosis [1].

2.1.4 Risk Factors

The following risk factors can cause hypertension: Smoking, Obesity, Stress level Consumption of alcohol, Sex, Consumption of salt, Genetic factors, Lack of exercise, Age [1].

2.1.5 Home Blood Pressure Monitoring

Using a home blood pressure monitor, we can monitor the patient and it is a practical way to see what the patient's blood pressure is like in their daily life. In order to achieve accurate readings, it is important to use the device correctly with the proper techniques.

2.1.6 Ambulatory Blood Pressure Monitoring (ABPM)

Ambulatory blood pressure monitoring (ABPM) is when the persons are on the move on a daily basis living their daily lives and blood pressure is being measured over time. Typically, this monitoring is done for 24 h using a special blood pressure device that is snug around the body and is connected to a cuff around your upper arm. The device is adequately sized so that it does not affect the daily life of the analyzed patient.

2.2 Computational Intelligence Techniques

2.2.1 Genetic Algorithms

A genetic algorithm is a search method that mimics Darwin's theory of biological evolution for problem solving. To do this, it is part of an initial population from which the most qualified individuals are selected for later reproduce them and mutate them to finally get the next generation of individuals that will be more adapted than the previous generation [2].

2.2.2 Chicken Swarm Optimization

Given the aforementioned descriptions, we can mathematically define CSO. For simplicity, we idealized the chickens' behaviors by the following rules. (1) In the chicken swarm, there exist several groups. Each group comprises a dominant rooster, a couple of hens, and chicks. (2) How to divide the chicken swarm into several groups and determine the identity of the chickens (roosters, hens and chicks) all depend on the fitness values of the chickens themselves. The chickens with best several fitness values would be acted as roosters, each of which would be the head rooster in a group. The chickens with worst several fitness values would be designated as chicks. The others would be the hens. The hens randomly choose which group to live in. The mother-child relationship between the hens and the chicks is also randomly established. (3) The hierarchal order, dominance relationship and mother-child relationship in a group will remain unchanged. These statuses only update every several (G) time steps. (4) Chickens follow their group-mate rooster to search for food, while they may prevent the ones from eating their own food.

Assume that RN, HN, CN and MN, indicate the number of the roosters, the hens, the chicks and the mother hens, respectively. The best RN chickens would be assumed to be roosters, while the worst CN ones would be regarded as chicks. The rest are treated as hens. All N virtual chickens, depicted by their positions $x_{i,j}^t$ ($i \in [1, \ldots, N]$, $j \in [1, \ldots, D]$), at time step t, search for food in a D-dimensional space. In this work, the optimization problems are minimization ones. Thus the best RN chickens correspond to the ones with RN minimal fitness values. Figure 2.1 shows the chicken swarm optimization framework [3].

Chicken Swarm Optimization. Framework of the CSO

Initialize a population of N chickens and define the related parameters;
Evaluate the N chickens' fitness values, $t=0$;
While ($t <$ Max_Generation)
   If ($t \% G == 0$)
     Rank the chickens' fitness values and establish a hierarchal order in the swarm;
     Divide the swarm into different groups, and determine the relationship between the chicks and mother hens in a group; End if
   For $i = 1 : N$
     If $i ==$ rooster Update its solution   End if
     If $i ==$ hen   Update its solution   End if
      If $i ==$ chick   Update its solution   End if
     Evaluate the new solution;
     If the new solution is better than its previous one, update it;
   End for
End while

Fig. 2.1 Chicken swarm optimization framework

2.2.3 Neural Networks

Neural networks are one of the most important recent scientific advances in knowledge. Through this tool we have tried to simulate one of the most important features that the human brain has, which is the ability to learn. This technique has been used to solve both scientific and daily life problems. With a good use of this technique, a wider technological development is possible in various fields, ranging from industry to medicine.

An artificial neural network is a system composed of many simple processing elements connected in parallel, whose function is determined by the structure of the network, the strength in the connections and the processing performed by the elements in the nodes. Neural networks just like people learn from experience.

One of the first network models was the Perceptron designed by Rosenblatt, which had three types of neurons: sensory, associative and response. The sensory took inputs from outside the network, the response units carried signals outside the network to the external world and the associative were exclusively internal. These types are now called output and hidden input units. Rosenblatt developed methods for the network to alter synaptic levels so that the network learned to recognize input levels.

After the 80s evolutionary computing became more popular and increased its scope. Algorithms have been developed to model high-level mental processes such as the association of concepts, deduction, induction and reasoning.

In 1986 Rumelhart and McClelland demonstrated that some impossible problems for simple Perceptrons can be solved by multi-level networks with non-linear activation functions, using simple training processes (back-propagation algorithm) [4].

2.2.3.1 Artificial Neural Networks

Artificial neural networks are inspired by the architecture of the nervous biological system, which consists of a large number of relatively simple neurons that work in parallel to facilitate rapid decision-making.

A neural network is a system of parallel processors connected to each other in the form of a directed graph. Schematically each processing element (neurons) of the network is represented as a node. These connections establish a hierarchical structure that, trying to emulate the physiology of the brain, seeks new processing models to solve specific real-world problems. The important thing in the development of the RNA technique is its useful behavior when learning, recognizing and applying relationships between objects and frames of real-world objects. In this sense, Artificial Neural Networks have been applied to a large number of real problems of considerable complexity. Its most important advantage is in solving problems that are too complex for conventional technologies, problems that do not have a solution algorithm or that their solution algorithm is very difficult to find [4].

2.2.4   Fuzzy Logic

The fuzzy or multi-valued logic is based on the fuzzy set theory proposed by Zadeh, which helps us in the modeling of knowledge, by means of fuzzy if-then rules [5–8].

2.2.4.1   Introduction to Fuzzy Logic

The human brain interprets inaccurate and incomplete sensory information provided by the perceptual organs. The fuzzy set theory provides a systematic calculation to deal with linguistic information, and that improves numerical computation through the use of linguistic tags stipulated by membership functions. On the other hand, a selection of fuzzy rules if-then forms the key component of a fuzzy inference system (FIS) that can effectively model human experience in a specific application.

2.2.4.2   Fuzzy Perception

Our interaction with the environment gives way to a perception of it, which can be considered fuzzy due to the way we express ourselves or assign linguistic values in everyday situations. Natural language handles vague and inaccurate concepts, for example, to verify the temperature of the water in a shower when opening and closing the hot and cold water faucets, we do not do so with the precision of a measuring instrument, rather we refer to it as cold, warm, hot, very hot, etc., and these characteristics give way to the use of also fuzzy rules that are of the "if this, then that" style where the antecedent and consequent of the rule can be a linguistic variable and the perception may be different for each individual.

2.2.4.3   Fuzzy Sets

Fuzzy sets are an extension of classical set theory and as the name implies, it is a set without well-established limits. That is, the transition from "belonging to a set" to "not belonging to a set" is gradual, and this smooth transition is characterized by degrees of membership that give diff sets flexibility in modeling commonly used linguistic expressions, such as "The weather is cold" or "Gustavo is high".

References

1. Mancia, G., Grassi, G., & Kjeldsen, S. E. (2008). *Manual of hypertension of the European society of hypertension.* Informa Healtcare: London, UK.

2. Guzmán, J. C., Melin, P., & Prado-Arechiga, G. (2015). Design of a fuzzy system for diagnosis of hypertension. In *Design of intelligent systems based on fuzzy logic, neural networks and nature-inspired optimization* (pp. 517–526). Springer International Publishing.
3. Meng X., Liu Y., Gao X., & Zhang H. (2014). A new bio-inspired algorithm: Chicken swarm optimization. In: Y. Tan, Y. Shi, C. A. C. Coello (Eds.), *Advances in swarm intelligence.* ICSI 2014. Lecture Notes in Computer Science (Vol. 8794). Cham: Springer.
4. Pulido, M., Mancilla, A., & Melin, P. (2010). Ensemble neural networks with fuzzy logic integration for complex time series prediction. *International Journal of Intelligent Engineering Informatics, 1*(1), 89–103.
5. Melin, P., Prado-Arechiga, G., Miramontes, I., & Medina-Hernandez, M. (2016). Hybrid intelligent model based on modular neural network and fuzzy logic for hypertension risk diagnosis. *Journal of Hypertension, 34.*
6. Nadia, J., Diedan, R., & Suryana, Y. (2017). Plate recognition using backpropagation neural network and genetic algorithm. *Procedia Computer Science, 116,* 365–372.
7. Ben Ali, J., Hamdi, T., Fnaiech, N., Di Costanzo, V. Fnaiech, F., & Ginoux, J.-M. (2018). Continuous blood glucose level prediction of Type 1 Diabetes based on Artificial Neural Network. *Biocybernetics and Biomedical Engineering, 38*(4), 828–840.
8. Castro, J. R., Castillo, O., & Melin, P. (2007). An interval type-2 fuzzy logic toolbox for control applications. In *2007 IEEE international fuzzy systems conference* (pp. 1–6).

Chapter 3
Proposed Neuro Fuzzy Hybrid Model

The proposed method is based on the neuro-fuzzy model for classification in medical diagnosis, which in this work it is being applied to blood pressure classification, where it is determined whether a patient suffers from hypertension or not.

3.1 General and Specific Neuro Fuzzy Hybrid Models

The Neuro Fuzzy General Hybrid Model is shown in Fig. 3.1, and within this model we can find that we have the data module, which will have all the patient information to analyze, these data and once analyzed is the input information to the neural network, from which output data are obtained, these data will be inputs to the fuzzy system that must have been optimized with a bioinspired algorithm to achieve better results as a final diagnosis.

Figure 3.2 shows the specific neuro-fuzzy hybrid, which shows how the model works in a specific way. First we have the database block, in this block there are two databases, the first database is obtained locally with the help of an expert in cardiology, this database has 300 patients, of which has been able to analyze 80 patients with the fuzzy classifier for tests. Recently, a second database of Framigham studies was obtained, in which there are 4000 patient records, of which 500 patient records were taken manually to perform experiments with fuzzy classifiers.

Once we have a database, the following data is analized as inputs: 45 systolic data and 45 diastolic data and 45 pulse data are taken. These are the inputs for the modules of the neural network and subsequently a trend is obtained, which is the input for the first fuzzy classifier proposed and carried out in this book, which takes as inputs the systolic and diastolic tensions, to subsequently give a value of the blood pressure level classification. Finally, this information is evaluated in the graphical

P. Melin et al., *Neuro Fuzzy Hybrid Models for Classification in Medical Diagnosis*,
SpringerBriefs in Computational Intelligence,
https://doi.org/10.1007/978-3-030-60481-3_3

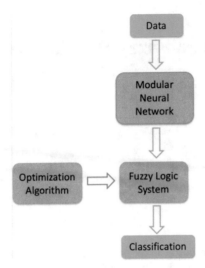

Fig. 3.1 General neuro fuzzy hybrid model

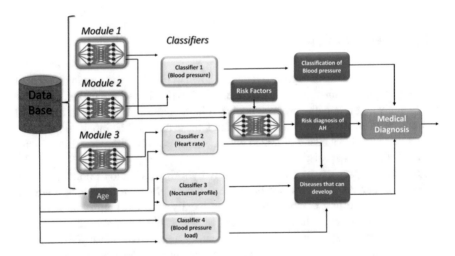

Fig. 3.2 Specific neuro fuzzy hybrid model

interface module corresponding to the blood pressure level. The second proposed classifier and marked with an orange border in Fig. 3.2 is the classification of blood pressure load, which obtains as input a percentage of pressure load and subsequently gives a level of blood pressure load, which is represented in the graphical interface module corresponding to this fuzzy classifier. The other modules shown in Fig. 3.2 are a complement to the general model, which works in a particular way and in a general way at the time of applying it in the final graphic interface. The modules corresponding to the fuzzy classifiers developed in this book are found in the appendix

Table 3.1 Architectures that were tested before choosing the optimal modular neural network

Architectures	Epoch	Layers	Neurons	Goal error	Learning rate	Mean systolic error	Mean diastolic error
Architecture 1	1000	3	10,10,5	0.0000001	0.01	0.242354	1.84563
Architecture 2	1000	3	10,5,5	0.00001	0.01	5.59638	2.44392
Architecture 3	1000	3	5,5,5	0.000001	0.01	6.77332	4.10991

section. This book is focusing on the fuzzy classifiers marked with the orange border, as indicated in Fig. 3.2.

3.2 Creation of the Modular Neural Network

Some experiments with different architectures were done with neural networks, these architecture examples are shown in Table 3.1 and the one with the best result was chosen according to the error given by the network. The modular neural network that was chosen has the following parameters: 2 modules, number of layers: 1 to 3, number of neurons: 1 to 20, for the modular neural network training: 1000 epochs, learning rate = 0.01, the error goal is of 0.0000001, 3 delays are considered with 70% of the data and Levenberg-Marquardt (trainlm) is the training method.

First, with the help of an expert, a 24 h monitoring was performed on 30 patients, in these experiments 45 blood pressure (BP) samples were obtained per patient, two tensions are considered which are: systolic and diastolic. This means that we have 45 systolic samples and 45 diastolic samples in a period of 24 h, approximately one sample every 20 min. This information is obtained with a 24-h monitoring device, which subsequently stores an excel file with the information collected during the 24 h.

Afterwards, the information of the excel file is analyzed, and the data is divided into 70% for training and 30% for testing. The modular neural network obtains the data as input, which has the systolic and diastolic modules, once the network models the information, it will produce a trend, which is tested with 30% of the data. The obtained systolic and diastolic trends are the inputs to the fuzzy classifier and these values are analyzed by the fuzzy system based on fuzzy rules and parameters of membership functions, which were designed based on the definition and classification of Blood pressure levels of the European Union as shown Table 2.1. The neural networks architectures based on previous tests are shown in Table 3.1. These architectures were obtained with the analyzed information of the patients, in this case, 4 samples were obtained per day, it is important to mention that it is working with a new neural network, which is optimized and receives 24-h monitoring, approximately 45 samples per patient to be able to give a better tendency to the fuzzy classifier.

Chapter 4
Study Cases to Test the Neuro Fuzzy Hybrid Model

4.1 Design of the Fuzzy Systems for Classification

The sections in this Chapter mention each of the classifiers that have been designed, the structure and parameters used for fuzzy systems, as well as the type of fuzzy system and number of fuzzy rules.

4.1.1 Design of the First Fuzzy Classifier for the Classification of Blood Pressure Levels

Throughout the tests performed with the Neuro Fuzzy Hybrid Model, the fuzzy classifier has been improved, which we initially had as a fuzzy system with two inputs, the systolic and diastolic pressures with eight membership functions for each and one output with eight membership functions, Mamdani type and 14 fuzzy rules.

The systolic input has a range from 20 to 300, in the diastolic input, the range is from 20 to 130 and the output is from 0 to 100. We have blood pressure levels such as Hypotension, Optimal, Normal, High Normal, Grade 1 hypertension, Grade 2 hypertension, Grade 3 hypertension and isolated systolic hypertension. The structure of first fuzzy classifier is shown in Fig. 4.1, the inputs for the classifier 1 are shown in Figs. 4.2 and 4.3 and finally Fig. 4.4 shows the output of the classifier 1.

© The Author(s), under exclusive license to Springer Nature Switzerland AG 2021   19
P. Melin et al., *Neuro Fuzzy Hybrid Models for Classification in Medical Diagnosis*,
SpringerBriefs in Computational Intelligence,
https://doi.org/10.1007/978-3-030-60481-3_4

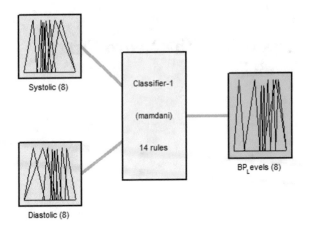

Fig. 4.1 Structure of the fuzzy logic classifier 1

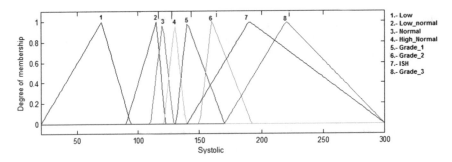

Fig. 4.2 Systolic input for the fuzzy logic classifier 1

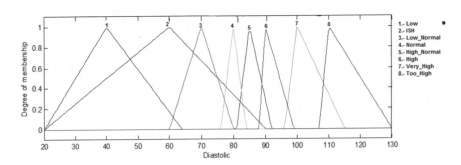

Fig. 4.3 Diastolic input for the fuzzy logic classifier 1

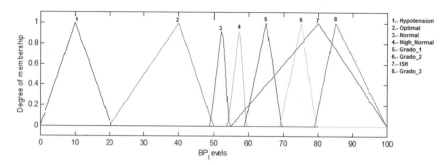

Fig. 4.4 BP_Levels is the output of the fuzzy logic classifier 1

4.1.2 Design of the Second Fuzzy Classifier for the Classification of Blood Pressure Levels

The second fuzzy system has two inputs, which are systolic and diastolic with seven membership functions for each and one output with ten membership functions, and a Mamdani type fuzzy system with 24 fuzzy rules based on an expert.

The systolic input has a range from 20 to 300, in the diastolic input the range is from 20 to 130 and at the output is from 0 to 100. We also have the blood pressure levels such as Hypotension, Optimal, Normal, High Normal, Grade 1 hypertension, Grade 2 hypertension, Grade 3 hypertension and isolated systolic hypertension Grade 1, isolated systolic hypertension Grade 2 and isolated systolic hypertension Grade 3. Figure 4.5 shows the structure of classifier 2, Figs. 4.6 and 4.7 show the inputs and Fig. 4.8 shows the output of the classifier 2.

Fig. 4.5 Structure of the fuzzy logic classifier 2

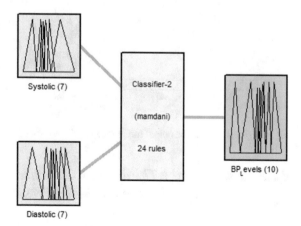

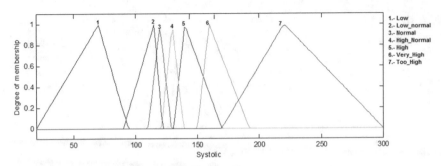

Fig. 4.6   Systolic input for the fuzzy logic classifier 2

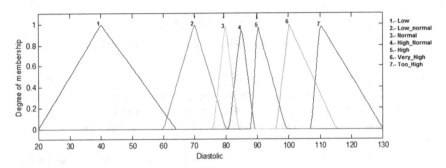

Fig. 4.7   Diastolic input for the fuzzy logic classifier 2

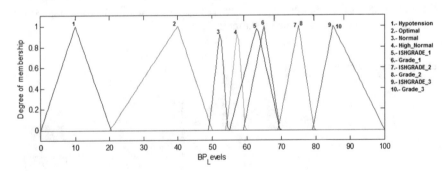

Fig. 4.8   BP_Levels is the output of the fuzzy logic classifier 2

4.1.3   *Design of the Third Fuzzy Classifier*
for the Classification of Blood Pressure Levels

In the third fuzzy system, it was decided to design it with the total number of possible
fuzzy rules based on the number of membership functions of the inputs and using
the product of this, and it was obtained a total of 49 fuzzy rules. The fuzzy system is

aimed observing how the classification is not good using the total of possible fuzzy rules and thus analyze the results obtained to later optimize this fuzzy system with bioinspired algorithms and find the optimal number of fuzzy rules to improve the results of the classifiers.

The number of fuzzy rules in a complete set of rules is equal to:

$$\text{TNPR} = \prod_{i=1}^{n} m_i \qquad (4.1)$$

where TNPR is the Total number of possible fuzzy rules; m_i, is the number of membership functions for input i and n is the number of inputs.

The third fuzzy classifier has two inputs systolic and diastolic pressures with seven membership functions for each and the output with ten membership functions, and Mamdani type with 49 fuzzy rules which are all possible.

The systolic input has a range from 20 to 300, in the diastolic input, the range is from 20 to 130 and at the output is from 0 to 100 and we have the blood pressure levels, such as Hypotension, Optimal, Normal, High Normal, Grade 1 hypertension, Grade 2 hypertension, Grade 3 hypertension and isolated systolic hypertension Grade 1, isolated systolic hypertension Grade 2 and isolated systolic hypertension Grade 3, we specify each of the ranges in the fuzzy rules. Figure 4.9 illustrates the structure of classifier 3, Figs. 4.10 and 4.11 show the inputs and Fig. 4.12 shows the output of the classifier 3.

Fig. 4.9 Structure of the fuzzy logic classifier 3

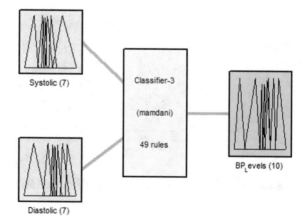

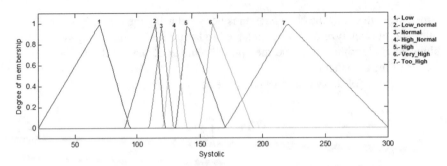

Fig. 4.10 Systolic input for the fuzzy logic classifier 3

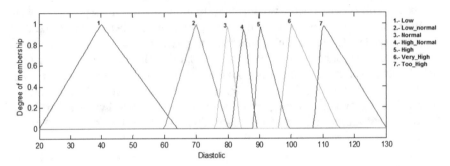

Fig. 4.11 Diastolic input for the fuzzy logic classifier 3

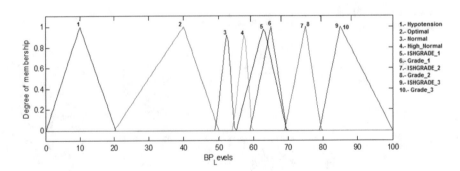

Fig. 4.12 BP_Levels is the output of the fuzzy logic classifier 3

4.1.4   The Optimization of the Fuzzy System Using a Genetic Algorithm (GA)

The classifier 4 was optimized with genetic algorithms, where we have a chromosome to optimize the fuzzy system, as shown in Fig. 4.13 and this chromosome has 122 genes, which help to optimize the structure of the fuzzy system, in this case, fuzzy

Systolic Input													
1	2	3	4	5	6	7	8	9	10	11	...		20
Diastolic Input													
21	22	23	24	25	26	27	28	29	30	31	...		41
BP_LEVELS Output													
42	43	44	45	46	47	48	49	50	51	52	...		72
Rules													
73	74	75	76	77	78	79	80	81	82	...		121	122

Fig. 4.13 Structure of the chromosome

Table 4.1 Parameters that were tested before choosing the optimal parameters for GA

Genetic algorithm	Generation	Population	Selection method	Mutation rate	Crossing rate
GA_1	100	100	Roulette wheel	0.06	0.5
GA_2	100	100	Roulette wheel	0.04	0.6
GA_3	100	100	Roulette wheel	0.06	0.7

rules and membership functions, Genes 1–72 (real numbers) allow to manage the parameters of the membership functions for inputs and output, genes 73–121 are the rules. The gene 122 allows reducing the number of rules, activating or deactivating them. The following Fig. 4.13 shows the structure of the chromosome [1, 2].

The algorithm parameters used are: generation: 100, population: 100, selection method: roulette wheel, crossing rate: 0.5, mutation rate: 0.06. The parameters are used since in previous tests a good error was obtained with these parameters.

The fitness function is based on the classification error as shown in Eq. (4.5), the idea is to minimize the classification error and this allows knowing that the base classifier is classifying in a correct way. The way to know if the classifier is classifying in a correct way is by using 2.1, which defines the blood pressure levels. Table 4.1 shows the different parameters used in the genetic algorithms.

4.1.5 Design of the Fuzzy Classifier Fourth Optimized with a GA

The fourth fuzzy classifier has two inputs, which are the systolic and diastolic pressures with seven membership functions each and the output with 10 membership functions, 21 optimized fuzzy rules, and Mamdani type of inference.

The systolic input has a range from 20 to 300, in the diastolic input, the range is from 20 to 130 and at the output is from 0 to 100 and we have blood pressure levels such as Hypotension, Optimal, Normal, High Normal, Grade 1 hypertension, Grade 2 hypertension, Grade 3 hypertension and isolated systolic hypertension Grade 1, isolated systolic hypertension Grade 2 and isolated systolic hypertension Grade 3.

Figure 4.14 illustrates the structure of the fourth fuzzy classifier, in the Figs. 4.15 and 4.16 shows the inputs and Fig. 4.17 shows the output of the classifier 4.

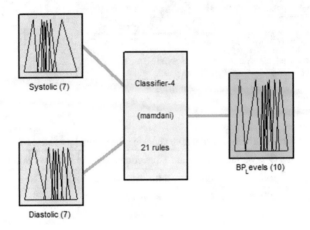

Fig. 4.14 Structure of the fuzzy logic classifier 4

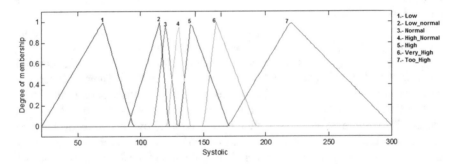

Fig. 4.15 Systolic input for the fuzzy logic classifier 4

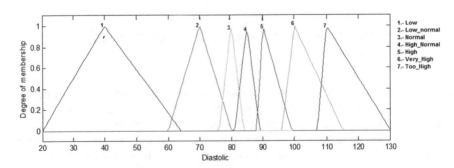

Fig. 4.16 Diastolic input for the fuzzy logic classifier 4

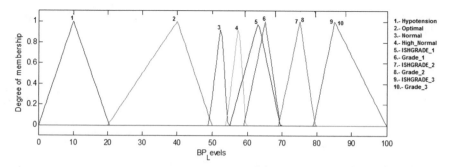

Fig. 4.17 BP_Levels is the output of the fuzzy logic classifier 4

4.1.6 Knowledge Representation of the Fuzzy Systems

In this section, we show how to represent the fuzzy system. The crisp output is calculated as follows: If the number of fired rules is **r** then the final blood pressure (BP) level is:

$$BP = \frac{\sum_{i=1}^{r} BP_i L_i}{\sum_{i=1}^{r} L_i} \tag{4.2}$$

where L_i is the firing level and BP_i is the crisp output of the if-th rule.

The triangular curve is a function of a vector, x, and depends on three scalar parameters a, b, and c, as given by

$$f(x; a, b, c) = \begin{cases} 0, & x \le a \\ \frac{x-a}{b-a}, & a \le x \le b \\ \frac{c-x}{c-b}, & b \le x \le c \\ 0, & c \le x \end{cases} \tag{4.3}$$

4.1.7 Results of the Proposed Method

The following tables show the results obtained for 30 patients, and based on these results we obtain the classification accuracy rate and classification error rate, for which we use the following equations:

The Classification Accuracy Rate (CA) is calculated as follows:

$$CA = \frac{N_c}{N_t} \tag{4.4}$$

where N_c is the number of training Instances correctly classified and N_t is the number of training instances.

The Classification Error Rate (CE) is calculated as follows:

$$CE = \frac{N_e}{N_t} \tag{4.5}$$

where N_e is the number of training instances incorrectly classified and N_t is the number of training instances.

The rows with bold letters are the incorrect classifications of each classifier, and in the following Table 4.2 we show the result of the first fuzzy logic system classifier.

We performed experiments using 24 h monitoring of 30 patients, from which the trend was obtained which was the one that entered to the fuzzy classifier, which produces the following results based on the accuracy rate in the classification of 30 patients. In this case, an accuracy rate of 80% was obtained that classified 24 of 30 patients correctly based on the result given by the ESH table and give us a classification error rate of 20%, which is equivalent to 6 of 30 incorrectly classified. Table 4.2 shows a summary of the results.

We conducted experiments using 24 h monitoring of 30 patients, from which the trend was obtained which was the one that entered to the fuzzy classifier, which produces the following results based on the accuracy rate in the classification of 30 patients, an accuracy rate of 90% was obtained, which classified 27 of 30 patients correctly based on the result given by the ESH table and a classification error rate of 10%, which is equivalent to 3 of 30 incorrectly classified as shows in Table 4.3. In the following Table 4.3 we show the result of the second fuzzy logic system classifier.

We carry out experiments using 24 h monitoring of 30 patients, from which the trend was obtained which was the one that entered to the fuzzy classifier, which gave us the following result based on the accuracy rate in the classification of 30 patients, an accuracy rate of 66.7% was obtained, which classified 20 of 30 patients correctly based on the result given by the ESH table and a classification error rate of 33.3%, which is equivalent to 10 of 30 incorrectly classified. In this case, Table 4.4 shows the accuracy rate was very bad result because the classifier was confused with many unnecessary rules; this is why we need to optimize the fuzzy rules to obtain the appropriate number of rules and obtain better results. In the following, Table 4.4 we show the result of the third fuzzy logic system classifier.

We conducted experiments of 30 patients using 24 h monitoring, from which the trend was obtained, which was the one that entered to the fuzzy classifier. The classifier gave us the following result based on the accuracy rate in the classification of 30 patients, an accuracy rate of 100% was obtained, which classified 30 of 30 patients correctly based on the result given by the ESH table and a classification error rate of 0%, which is equivalent to 0 of 30 incorrectly classified. In these experiments, Table 4.5 shows the results were very good since the classifier was successful in the total of the tests. In the following, Table 4.5 we show the result of the fourth fuzzy logic system classifier.

Table 4.2 Results of the 30 patients who were monitored and classified in the classifier 1

Patient	Systolic	Diastolic	Classifier 1	Fuzzy percentage	ESH BP_Levels table
1	**139**	**84**	**Normal**	**50**	**High normal**
2	135	90	Grade_1	62.3	Grade_1
3	160	98	Grade_2	72.3	Grade_2
4	177	110	Grade_3	84.6	Grade_3
5	142	85	ISH	77.5	Ish_grade_1
6	160	89	ISH	71.6	Ish_grade_2
7	182	89	ISH	82.1	Ish_grade_3
8	85	50	Hypotension	10.2	Hypotension
9	110	70	Optimal	36.6	Optimal
10	125	82	Normal	54	Normal
11	135	85	High Normal	57	High normal
12	**159**	**94**	**Grade_2**	**70.2**	**Grade 1**
13	175	105	Grade_2	79.3	Grade 2
14	180	110	Grade_3	84.2	Grade 3
15	**110**	**80**	**Optimal**	**50**	**Normal**
16	128	89	High normal	57	High normal
17	158	70	ISH	77.9	Ish_grade_1
18	**150**	**108**	**Grade_3**	**83.2**	**Grade_2**
19	199	95	Grade_3	88.6	Grade_3
20	**179**	**99**	**Grade_3**	**81.8**	**Grade_2**
21	181	100	Grade_3	82.8	Grade_3
22	210	90	Grade_3	88.1	Grade_3
23	140	100	Grade_2	74.5	Grade_2
24	159	120	Grade_3	88.5	Grade_3
25	178	115	Grade_3	88.6	Grade_3
26	**140**	**80**	**Normal**	**50**	**Ish grade 1**
27	150	89	ISH	68.2	Ish_grade_1
28	179	80	ISH	77.9	Ish_grade_2
29	179	89	ISH	80.6	Ish_grade_2
30	199	82	ISH	77.8	Ish_grade_3

Once the experiments were performed for each of the classifiers, a comparison was made among them. The fuzzy classifiers analyzed are First Classifier Type-1(C1T1), Second Classifier Type-1 (C2T1), Third Classifier Type-1 (C3T1), Fourth Classifier Optimized Type-1 (C4T1O), fifth classifier Type-2 (C5T2) and Sixth classifier Optimized Type-2 (C6T2O). Table 4.6 shows the results for 30 patients with each of the proposed classifiers and Tables 4.7 and 4.8 shows the results obtained for 60 and 80 patients respectively.

Table 4.3 Shows the results of the 30 patients who were monitored and classified in the classifier 2

Patient	Systolic	Diastolic	Classifier 2	Fuzzy percentage	ESH BP_Levels table
1	**139**	**84**	**Grade_1**	**62.4**	**High normal**
2	135	90	Grade_1	62.5	Grade_1
3	160	98	Grade_2	70.9	Grade_2
4	177	110	Grade_3	84.4	Grade_3
5	142	85	Ish_grade_1	61.8	Ish_grade_1
6	160	89	Grade_2	70.2	Ish_grade_2
7	**182**	**89**	**Grade_2**	**76.7**	**Ish_grade_3**
8	85	50	Hypotension	10.2	Hypotension
9	110	70	Optimal	36.6	Optimal
10	125	82	Normal	54.4	Normal
11	**135**	**85**	**Grade_1**	**61.1**	**High normal**
12	159	94	Grade_2	70	Grade_1
13	175	105	Grade_2	78.9	Grade_2
14	180	110	Grade_3	84.8	Grade_3
15	110	80	Normal	52	Normal
16	128	89	High normal	60.7	High normal
17	158	70	Ish_grade_1	70.5	Ish_grade_1
18	150	108	Grade_2	77.1	Grade_2
19	199	95	Grade_3	81.3	Grade_3
20	179	99	Grade_2	80.5	Grade_2
21	181	100	Grade_3	80.9	Grade_3
22	210	90	Grade_3	80	Grade_3
23	140	100	Grade_2	69.5	Grade_2
24	159	120	Grade_3	79.3	Grade_3
25	178	115	Grade_3	84.5	Grade_3
26	140	80	Ish_grade_1	60.2	Ish_grade_1
27	150	89	Ish_grade_1	64.5	Ish_grade_1
28	179	80	Ish_grade_2	73.3	Ish_grade_2
29	179	89	Ish_grade_2	75.4	Ish_grade_2
30	199	82	Ish_grade_3	78.9	Ish_grade_3

4.1.8 Comparison of Results

We compare classifier 2 that is based on an expert with classifier 4 that is optimized with the genetic algorithm, and it based on the total of possible rules based on the European hypertension guide, this is shown in 2.1.

Table 4.4 Results of the 30 patients who were monitored and classified in the classifier 3

Patient	Systolic	Diastolic	Classifier 3	Fuzzy percentage	ESH BP_Levels table
1	**139**	**84**	**Ish_grade_1**	**62.5**	**High normal**
2	135	90	Grade_1	64.4	Grade_1
3	160	98	Grade_2	72.3	Grade_2
4	177	110	Grade_3	84.6	Grade_3
5	142	85	Ish_grado_1	62.5	Ish_grade_1
6	**160**	**89**	**Grade_2**	**70.2**	**Ish_grade_2**
7	**182**	**89**	**Grade_2**	**83.4**	**Ish_grade_3**
8	85	50	Hypotension	10.2	Hypotension
9	110	70	Optimal	36.6	Optimal
10	125	82	Normal	54.5	Normal
11	**135**	**85**	**Grade_1**	**61.5**	**High normal**
12	**159**	**94**	**Grade_2**	**70.2**	**Grade_1**
13	175	105	Grade_2	79.3	Grade_2
14	**180**	**110**	**Grade_2**	**84.2**	Grade_3
15	110	80	Normal	52	Normal
16	**128**	**89**	**Grade_1**	**64.4**	**High normal**
17	**158**	**70**	**Grade_2**	**70.5**	Ish_grade_1
18	150	108	Grade_2	83.2	Grade_2
19	199	95	Grade_3	88.6	Grade_3
20	179	99	Grade_2	81.8	Grade_2
21	181	100	Grade_3	82.8	Grade_3
22	210	90	Grade_3	88.1	Grade_3
23	140	100	Grade_2	74.5	Grade_2
24	159	120	Grade_3	88.5	Grade_3
25	178	115	Grade_3	88.6	Grade_3
26	140	80	Ish_grado_1	62.5	Ish_grade_1
27	**150**	**89**	**Grade_1**	**64.4**	**Ish_grade_1**
28	179	80	Ish_grade_2	81.8	Ish_grade_2
29	**179**	**89**	**Grade_2**	**81.7**	**Ish_grade_2**
30	199	82	Ish_grade_3	88.5	Ish_grade_3

Of the 30 patients that were monitored, the classification achieved by classifier 2 with 24 rules given by an expert, the result was: accuracy rate of 90% with a 10% error, this is shown in Table 4.9 and in classifier 4 optimized with the new Computational method and reduced to 21 rules, gives as a result: 100% accuracy rate and 0% error rate, and this is shown in Table 4.10.

The following tables show the comparison of normal and optimized fuzzy classifiers; these experiments were performed with 30, 60 and 80 patients. Table 4.11

Table 4.5 Results of the 30 patients who were monitored and classified in the classifier 4

Patient	Systolic	Diastolic	Optimized classifier 4	Fuzzy percentage	ESH BP_Levels table
1	139	84	High normal	61.3	High normal
2	135	90	Grade_1	62.5	Grade_1
3	160	98	Grade_2	74	Grade_2
4	177	110	Grade_3	84.3	Grade_3
5	142	85	Ish_grade_1	61.3	Ish_grade_1
6	160	89	Ish_grade_2	71.8	Ish_grade_2
7	182	89	Ish_grade_3	83.2	Ish_grade_3
8	85	50	Hypotension	10.2	Hypotension
9	110	70	Optimal	36.6	Optimal
10	125	82	Normal	55.2	Normal
11	135	85	High normal	60.8	High normal
12	159	94	Grade_1	71.8	Grade_1
13	175	105	Grade_2	79.3	Grade_2
14	180	110	Grade_3	84	Grade_3
15	110	80	Normal	52	Normal
16	128	89	High normal	56.9	High normal
17	158	77	Ish_grade_1	66.4	Ish_grade_1
18	150	108	Grade_2	82.9	Grade_2
19	199	95	Grade_3	87.8	Grade_3
20	179	99	Grade_2	81.6	Grade_2
21	181	100	Grade_3	82.6	Grade_3
22	210	90	Grade_3	87.4	Grade_3
23	140	100	Grade_2	75.8	Grade_2
24	159	120	Grade_3	87.7	Grade_3
25	178	115	Grade_3	87.8	Grade_3
26	140	80	Ish_grade_1	59.7	Ish_grade_1
27	150	89	Ish_grade_1	65.2	Ish_grade_1
28	179	80	Ish_grade_2	73.8	Ish_grade_2
29	179	89	Ish_grade_2	81.6	Ish_grade_2
30	199	82	Ish_grade_3	77.8	Ish_grade_3

shows the comparison of normal and optimized type-1 classifier with 30 patients. Tables 4.12 and 4.13 show the result obtained with 60 and 80 patients respectively.

Table 4.6 Summary of the results for the different fuzzy classifiers with 30 patients

Results for 30 patients

	C1T1	C2T1	C3T1	C4T1O	C5T2	C6T2O
Patients	30	30	30	30	30	30
Accuracy rate (%)	80	90	66.70	100	100	100
Correct classification	24	27	20	30	30	30
Error rate (%)	20	10	33.30	0	0	0
Total errors	6	3	10	0	0	0

Table 4.7 Summary of the results for the different fuzzy classifiers with 60 patients

Results for 60 patients

	C1T1	C2T1	C3T1	C4T1O	C5T2	C6T2O
Patients	60	60	60	60	60	60
Accuracy rate (%)	78.33	91.67	63.33	100	91.67	100
Correct classification	47	55	38	60	55	60
Error rate (%)	21.67	8.33	36.67	0	8.33	0
Total errors	13	5	22	0	5	0

Table 4.8 Summary of the results for the different fuzzy classifiers with 80 patients

Results for 80 patients

	C1T1	C2T1	C3T1	C4T1O	C5T2	C6T2O
Patients	80	80	80	80	80	80
Accuracy rate (%)	77.50	86.25	62.50	96.25	86.25	96.25
Correct classification	62	69	50	77	69	77
Error rate (%)	22.50	13.75	37.50	3.75	13.75	3.75
Total errors	18	11	30	3	11	3

4.2 A Comparative Study Between European Guidelines and American Guidelines Using Fuzzy Systems for the Classification of Blood Pressure

The use of soft computing tools for the diagnosis of blood pressure are increasingly important, these tools help the rapid and accurate diagnosis of a patient. This is why the use of fuzzy classifiers, which allow handling uncertainty in the information, and can make a quick and accurate diagnosis are important. For this reason, based on this study we want to identify the way in which a patient is diagnosed using two guides of different blood pressure levels, the first is the European guidelines and the second it is the American guides, the latter mentioned was recently updated and it is

Table 4.9 Results of the 30 patients who were monitored and classified in the classifier based in an expert

Patient	Systolic	Diastolic	Classifier 2	Fuzzy percentage	ESH BP_Levels table
1	**139**	**84**	**Grade 1**	**62.4**	**High normal**
2	135	90	Grade 1	62.5	Grade 1
3	160	98	Grade 2	70.9	Grade 2
4	177	110	Grade 3	84.4	Grade 3
5	142	85	ish_grado1	61.8	Ish grade 1
6	160	89	Grade 2	70.2	Ish_Grade 2
7	**182**	**89**	**Grade 2**	**76.7**	**Ish_Grade 3**
8	85	50	Hypotension	10.2	Hypotension
9	110	70	Optimal	36.6	Optimal
10	125	82	Normal	54.4	Normal
11	**135**	**85**	**Grade 1**	**61.1**	**High normal**
12	159	94	Grade 2	70	Grade 1
13	175	105	Grade 2	78.9	Grade 2
14	180	110	Grade 3	84.8	Grade 3
15	110	80	Normal	52	Normal
16	128	89	High Normal	60.7	High normal
17	158	70	Ish grade 1	70.5	Ish grade 1
18	150	108	Grade 2	77.1	Grade 2
19	199	95	Grade 3	81.3	Grade 3
20	179	99	Grade 2	80.5	Grade 2
21	181	100	Grade 3	80.9	Grade 3
22	210	90	Grade 3	80	Grade 3
23	140	100	Grade 2	69.5	Grade 2
24	159	120	Grade 3	79.3	Grade 3
25	178	115	Grade 3	84.5	Grade 3
26	140	80	Ish grade 1	60.2	Ish grade 1
27	150	89	Ish grade 1	64.5	Ish grade 1
28	179	80	Ish grade 2	73.3	Ish grade 2
29	179	89	Ish grade 2	75.4	Ish grade 2
30	199	82	Ish grade 3	78.9	Ish grade 3

important to see the impact that each of them can have on society when classifying a patient.

The main objective of using fuzzy classifiers is to provide accuracy in the handling of information, which will help to be more accurate when classifying the blood pressure level of a patient, these are based on the parameters provided by the guidelines used in this study, and make the comparison with the levels given by the European

Table 4.10 Results of the 30 patients that were monitored and classified with the classifier optimized with GA

Patient	Systolic	Diastolic	Optimized classifier 4	Fuzzy percentage	ESH BP_Levels table
1	139	84	High normal	61.3	High normal
2	135	90	Grade 1	62.5	Grade 1
3	160	98	Grade 2	74	Grade 2
4	177	110	Grade 3	84.3	Grade 3
5	142	85	Ish grade 1	61.3	Ish grade 1
6	160	89	Ish grade 2	71.8	Ish_grade 2
7	182	89	Ish grade 3	83.2	Ish_grade 3
8	85	50	Hypotension	10.2	Hypotension
9	110	70	Optimal	36.6	Optimal
10	125	82	Normal	55.2	Normal
11	135	85	High normal	60.8	High normal
12	159	94	Grade 1	71.8	Grade 1
13	175	105	Grade 2	79.3	Grade 2
14	180	110	Grade 3	84	Grade 3
15	110	80	Normal	52	Normal
16	128	89	High normal	56.9	High normal
17	158	77	Ish grade1	66.4	Ish grade 1
18	150	108	Grade 2	82.9	Grade 2
19	199	95	Grade 3	87.8	Grade 3
20	179	99	Grade 2	81.6	Grade 2
21	181	100	Grade 3	82.6	Grade 3
22	210	90	Grade 3	87.4	Grade 3
23	140	100	Grade 2	75.8	Grade 2
24	159	120	Grade 3	87.7	Grade 3
25	178	115	Grade 3	87.8	Grade 3
26	140	80	Ish grade 1	59.7	Ish grade 1
27	150	89	Ish grade 1	65.2	Ish grade 1
28	179	80	Ish grade 2	73.8	Ish grade 2
29	179	89	Ish grade 2	81.6	Ish grade 2
30	199	82	Ish grade 3	77.8	Ish grade 3

guide and the levels given by the American guide and observe the behavior that each of them can take at the time of making a diagnosis and see the blood pressure level given to the patient analyzed, which can generate a significant difference in people who are currently not diagnosed hypertensive using European guidelines, but if they

Table 4.11 Comparison of normal and optimized type-1 classifier with 30 patients

Comparison of normal and optimized type-1 classifier

	C2T1	C4T1O
Patients	30	30
Accuracy rate (%)	90	100
Error rate	10	0

Comparison of normal and optimized type-2 classifier

	C5T2	C6T2O
Patients	30	30
Accuracy rate (%)	100	100
Error rate (%)	0	0

Table 4.12 Comparison of normal and optimized type-1 classifier with 60 patients

Comparison of normal and optimized type-1 classifier

	C2T1	C4T1O
Patients	60	60
Accuracy rate (%)	91.67	100
Error rate (%)	8.33	0

Comparison of normal and optimized type-2 classifier

	C5T2	C6T2O
Patients	60	60
Accuracy rate (%)	91.67	100
Error rate (%)	8.33	0

Table 4.13 Shows the comparison of normal and optimized type-1 classifier with 80 patients

Comparison of normal and optimized type-1 classifier

	C2T1	C4T1O
Patients	80	80
Accuracy rate(%)	86.25	96.25
Error rate(%)	13.75	3.75

Comparison of normal and optimized type-2 classifier

	C5T2	C6T2O
Patients	80	80
Accuracy rate (%)	86.25	96.25
Error rate (%)	13.75	3.75

are hypertensive using the American guidelines which influences having to change their lifestyle.

Currently there is a database with 200 patients, each patient has an average 45 systolic measurements and 45 diastolic measurements, these 24-h screenings have been obtained in collaboration with a cardiologist. This information is then processed with neural networks and obtains the tendency, this tendency enters in the fuzzy classifier, which gives us the blood pressure level depending on the base guide, which can be the American or the European guide. In this study, we analyzed 30 randomly selected patients from the 200 patients in our database.

4.2.1 Experiments and Results

Of the 30 patients selected for the experiments, the following result was obtained based on the parameters and levels given by each of the guidelines and using fuzzy classifiers for each of them, which have a correct classification accuracy rate of 100% classification for the 30 patients processed and classified. For the European guidelines, the classification of patients with hypertension is 43.3% with a standard deviation of 2.71 and using the new American guidelines is 56.6% was obtained with a standard deviation of 2.71 using the same patients for each of them. Table 4.14 shows the comparison of patients suffering from hypertension using different guidelines. Table 4.15 shows the comparison of results obtained with the guides analyzed.

The study conducted resulted in a higher rate of hypertensive people, which based on European guidelines are within normal to high normal ranges, but based on American guidelines it was observed that some patients directly enter hypertensive: stage 1 or stage 2, which generates an impact on their daily life, in which they need to change their lifestyle to avoid a cardiovascular event.

4.3 Optimal Genetic Design of Type-1 and Interval Type-2 Fuzzy Systems for Blood Pressure Level Classification

In this particular case, a fuzzy classifier of Mamdani type was built, with 21 rules, with two inputs and one output and the objective of this classifier is to perform blood

Table 4.14 Comparison of European guidelines and American guidelines patients with hypertension

	European guidelines	American guidelines
Patients	30	30
% of patients with hypertension	43.30	56.60

Table 4.15 Comparison between the different guides and the fuzzy classifiers

Systolic	Diastolic	European Guidelines	European Fuzzy classifier	American Guidelines	American Fuzzy classifier
151	89	ISH 1	ISH 1	Stage 2	Stage 2
143	96	Grade 1	Grade 1 Hypertension	Stage 2	Stage 2
134	61	High_Normal	Normal	Stage 1	Stage 1
121	78	Normal	Normal	Elevated	Elevated
108	64	Optimal	Optimal	Normal	Normal
162	86	ISH 2	ISH 2	Stage 2	Stage 2
130	86	High_Normal	High_Normal	Stage 1	Stage 1
116	73	Optimal	Optimal	Normal	Normal
118	57	Optimal	Optimal	Normal	Normal
158	80	ISH 1	ISH 1	Stage 2	Stage 2
160	94	Grade 2	ISH 2	Stage 2	Stage 2
124	82	Normal	Normal	Stage 1	Stage 1
104	63	Optimal	Optimal	Normal	Normal
121	70	Normal	Normal	Elevated	Elevated
152	106	Grade 2	Grade 2 Hypertension	Stage 2	Stage 2
133	101	Grade 2	High_Normal	Stage 2	Stage 2
128	61	Normal	Optimal	Elevated	Elevated
113	69	Optimal	Optimal	Normal	Normal
154	95	Grade 1	Grade 1 Hypertension	Stage 2	Stage 2
159	76	ISH 1	ISH 1	Stage 2	Stage 2
108	68	Optimal	Optimal	Normal	Normal
126	64	Normal	Normal	Elevated	Elevated
161	103	Grade 2	Grade 2 Hypertension	Stage 2	Stage 2
130	77	High_Normal	High_Normal	Stage 1	Stage 1
96	61	Optimal	Optimal	Normal	Normal
143	94	Grade 1	Grade 1 Hypertension	Stage 2	Stage 2
117	60	Optimal	Optimal	Normal	Normal
126	68	Normal	Normal	Elevated	Elevated
149	100	ISH_2	ISH 2	Stage 2	Stage 2
140	81	ISH 1	ISH 1	Stage 2	Stage 2
114	74	Optimal	Optimal	Normal	Normal
151	87	ISH 1	ISH 1	Stage 2	Stage 2
121	75	Normal	Normal	Elevated	Elevated
130	93	Grade 1	Grade 1 Hypertension	Stage 2	Stage 2
115	75	Optimal	Optimal	Normal	Normal
124	74	Normal	Normal	Elevated	Elevated

pressure level classification based on knowledge of an expert, which is represented in the fuzzy rules. Subsequently different architectures were made in type-1 and type-2 fuzzy systems for classification, where the parameters of the membership functions used in the design of each architecture were adjusted, which can be triangular, trapezoidal and Gaussian, as well as how the fuzzy rules are optimized based on the ranges established by an expert.

4.3.1 Design of the Type-1 Fuzzy Systems for Classification with Triangular Membership Functions

The design of this type-1 fuzzy system was made based on previous work [3] where the membership functions and the fuzzy rules are optimized to find the best possible classification architecture, after different experiments were performed it was obtained that the architecture with triangular membership functions produced better results when using type-1 fuzzy systems. The design is similar to works done in other application areas [4].

4.3.2 Design of the Type-1 FS for Classification with Trapezoidal Membership Functions

The following sections specify each of the classifiers that have been designed with type-1 fuzzy systems and trapezoidal membership functions, the structure of the fuzzy system and the parameters used for each of them, as well as the number of fuzzy rules and fuzzy system type are presented.

4.3.2.1 Design of the Second FS for the Classification of BP Levels with Trapezoidal Memberships Functions

The structure of the fuzzy system is shown in Fig. 4.18. The numbers marked in the Fig. 4.19 list each of the MFs for the input systolic and these are: 1—Low, 2—Low_Normal, 3—Normal, 4—High_Normal, 5—High, 6—Very_High, 7—Too_High.

The numbers marked in Fig. 4.20 list each of the membership functions for the input diastolic and these are: 1—Low, 2—Low_Normal, 3—Normal, 4—High_Normal, 5—High, 6—Very_High, 7—Too_High.

The numbers marked in Fig. 4.21 list each of the membership functions for the output BP_Levels and these are: 1—Hypotension, 2—Optimal, 3—Normal, 4—High_Normal, 5—ISH_GRADE_1, 6—Grade_1, 7—ISH_GRADE_2, 8—Grade_2, 9—ISH_GRADE_3, 10—Grade_3.

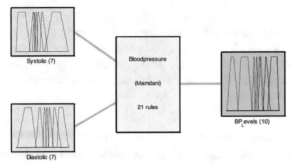

System Bloodpressure: 2 inputs, 1 outputs, 21 rules

Fig. 4.18 Structure of the type-1 FS for classification with trapezoidal membership functions

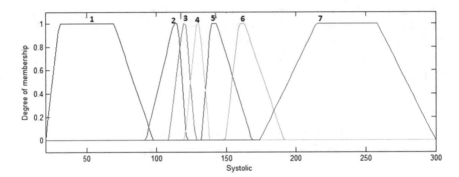

Fig. 4.19 Systolic input for the type-1 FS for classification with trapezoidal membership functions

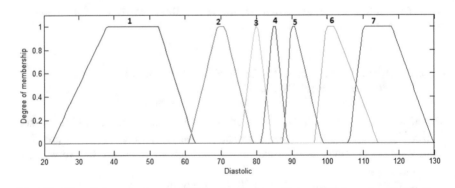

Fig. 4.20 Diastolic input for the type-1 FS for classification with trapezoidal membership functions

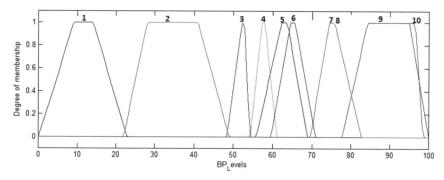

Fig. 4.21 BP_Levels output for the type-1 FS for classification with trapezoidal membership functions

Systolic Input																											
Low				Low_Normal				Normal				High_Normal				High				Very_High				Too_High			
1	2	3	4	5	6	7	8	9	10	11	12	13	14	15	16	17	18	19	20	21	22	23	24	25	26	27	28
Diastolic Input																											
Low				Low_Normal				Normal				High_Normal				High				Very_High				Too_High			
29	30	31	32	33	34	35	36	37	38	39	40	41	42	43	44	45	46	47	48	49	50	51	52	53	54	55	56
BP_Levels Output																											
Hypotension				Optimal				Normal				High_Normal				ISH_Grade_1				Grade_1				ISH_Grade_2			
57	58	59	60	61	62	63	64	65	66	67	68	69	70	71	72	73	74	75	76	77	78	79	80	81	82	83	84
BP_Levels Output																											
Grade_2				ISH_Grade_3				Grade_3																			
85	86	87	88	89	90	91	92	93	94	95	96																

Fig. 4.22 Structure of the chromosome for the type-1 FS for classification with trapezoidal MFs

4.3.2.2 Genetic Type-1 Fuzzy System with Trapezoidal Membership Functions

The fuzzy system (FS) was optimized with GA, in the GA it is necessary a chromosome to optimize the (MFs), as shown in Fig. 4.22 and the chromosome has 96 genes and this help to optimize the MFs. In this case, Genes 1–28 (real numbers) allow to manage the parameters of the systolic input, Genes 29–56 (real numbers) allow to manage the parameters of the diastolic input and Genes 57–96 (real numbers) allow to manage the parameters of the BP_Levels output. The following Fig. 4.22 shows the structure of the chromosome.

4.3.2.3 Design of the Optimized Type-1 FS for Classification with Trapezoidal Membership Functions

The structure of the optimized type-1 fuzzy system is shown in Fig. 4.23. The numbers marked in the Fig. 4.24 list each of the membership functions for the input systolic

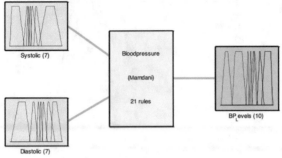

System Bloodpressure: 2 inputs, 1 outputs, 21 rules

Fig. 4.23 Structure of the optimized type-1 FS for classification with trapezoidal membership functions

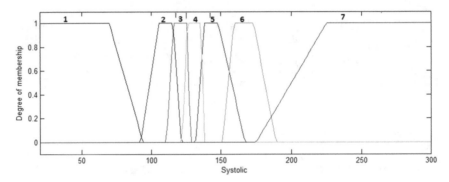

Fig. 4.24 Systolic input for the optimized type-1 FS for classification with trapezoidal membership functions

and these are: 1—Low, 2—Low_Normal, 3—Normal, 4—High_Normal, 5—High, 6—Very_High, 7—Too_High.

The numbers marked in Fig. 4.25 list each of the membership functions for the input diastolic and these are: 1—Low, 2—Low_Normal, 3—Normal, 4—High_Normal, 5—High, 6—Very_High, 7—Too_High.

The numbers marked in Fig. 4.26 list each of the membership functions for the output BP_Levels and these are: 1—Hypotension, 2—Optimal, 3—Normal, 4—High_Normal, 5—ISH_GRADE_1, 6—Grade_1, 7—ISH_GRADE_2, 8—Grade_2, 9—ISH_GRADE_3, 10—Grade_3.

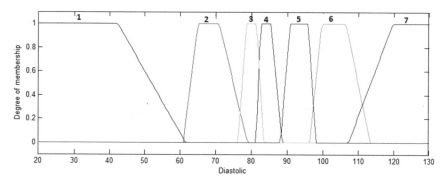

Fig. 4.25 Diastolic input for the optimized type-1 FS for classification with trapezoidal membership functions

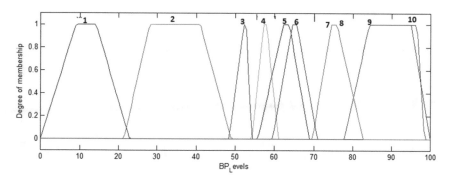

Fig. 4.26 BP_Levels output for the optimized type-1 FS for classification with trapezoidal membership functions

4.3.3 Design of the Type-1 FS for Classification with Gaussian Membership Functions

The following sections specify each of the classifiers that have been performed with type-1 fuzzy systems and Gaussian membership functions, the structure of the fuzzy system and the parameters used for each of them, as well as the number of fuzzy rules and fuzzy system type.

4.3.3.1 Design of the Third FS for the Classification of BP Levels with Gaussian Membership Functions

The structure of the fuzzy system is shown in Fig. 4.27. The numbers marked in Fig. 4.28 list each of the membership functions for the input systolic and these are: 1—Low, 2—Low_Normal, 3—Normal, 4—High_Normal, 5—High, 6—Very_High, 7—Too_High.

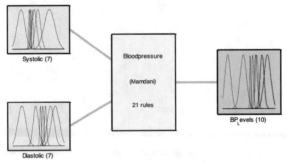

Fig. 4.27 Structure of the type-1 fuzzy system for classification with Gaussian membership functions

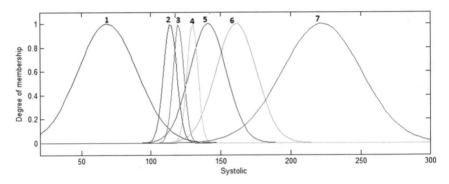

Fig. 4.28 Systolic input for the type-1 FS for classification with Gaussian membership functions

The numbers marked in Fig. 4.29 list each of the membership functions for the input diastolic and these are: 1—Low, 2—Low_Normal, 3—Normal, 4—High_Normal, 5—High, 6—Very_High, 7—Too_High.

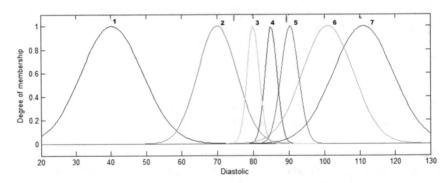

Fig. 4.29 Diastolic input for the type-1 fuzzy system for classification with Gaussian membership functions

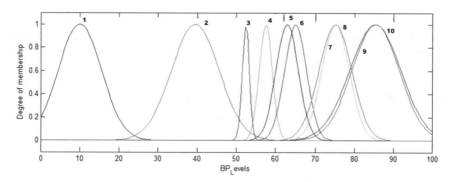

Fig. 4.30 BP_Levels output for the type-1 fuzzy system for classification with Gaussian membership functions

The numbers marked in Fig. 4.30 list each of the membership functions for the output BP_Levels and these are: 1—Hypotension, 2—Optimal, 3—Normal, 4—High_Normal, 5—ISH_GRADE_1, 6—Grade_1, 7—ISH_GRADE_2, 8—Grade_2, 9—ISH_GRADE_3, 10—Grade_3.

4.3.3.2 Genetic Type-1 Fuzzy System with Gaussian Membership Functions

The fuzzy system was optimized with GA, in the GA it is necessary to use a chromosome to optimize the membership functions (MFS), as shown in Fig. 4.31 and the chromosome has 48 genes and this information help to optimize the membership functions. In this case, Genes 1–14 (real numbers) allow to manage the parameters of the systolic input, Genes 15–28 (real numbers) allow to manage the parameters

Systolic Input													
Low		Low_Normal		Normal		High_Normal		High		Very_High		Too_High	
1	2	3	4	5	6	7	8	9	10	11	12	13	14
Diastolic Input													
Low		Low_Normal		Normal		High_Normal		High		Very_High		Too_High	
15	16	17	18	19	20	21	22	23	24	25	26	27	28
BP_Levels Output													
Hypotension		Optimal		Normal		High_Normal		ISH_Grade_1		Grade_1		ISH_Grade_2	
29	30	31	32	33	34	35	36	37	38	39	40	41	42
Grade_2		ISH_Grade_3		Grade_3									
43	44	45	46	47	48								

Fig. 4.31 Structure of the chromosome for the type-1 FS for classification with Gaussians membership functions

of the diastolic input and Genes 29–48 (real numbers) allow to manage the parameters of the BP_Levels output. The following Fig. 4.31 shows the structure of the chromosome.

4.3.3.3 Design of the Optimized Type-1 FS for Classification with Gaussians Membership Functions

The structure of the fuzzy system with two inputs and one output is shown in Fig. 4.32. The numbers marked in the Fig. 4.33 list each of the membership functions for the input systolic and these are: 1—Low, 2—Low_Normal, 3—Normal, 4—High_Normal, 5—High, 6—Very_High, 7—Too_High.

The numbers marked in Fig. 4.34 list each of the membership functions for the input diastolic and these are: 1—Low, 2—Low_Normal, 3—Normal, 4—High_Normal, 5—High, 6—Very_High, 7—Too_High.

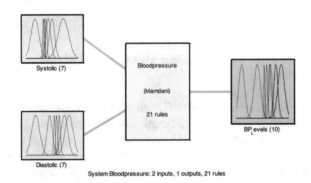

Fig. 4.32 Structure of the optimized type-1 FS for classification with Gaussians membership functions

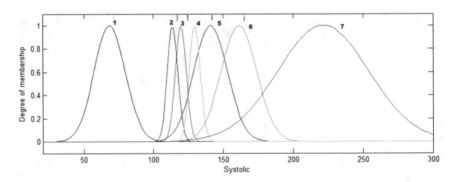

Fig. 4.33 Systolic input for the optimized type-1 FS for classification with Gaussians membership functions

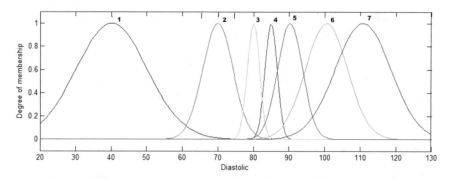

Fig. 4.34 Diastolic input for the optimized type-1 FS for classification with Gaussian membership functions

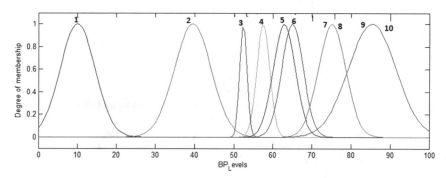

Fig. 4.35 BP_Levels output for the optimized type-1 FS for classification with Gaussians membership functions

The numbers marked in Fig. 4.35 list each of the membership functions for the output BP_Levels and these are: 1—Hypotension, 2—Optimal, 3—Normal, 4—High_Normal, 5—ISH_GRADE_1, 6—Grade_1, 7—ISH_GRADE_2, 8—Grade_2, 9—ISH_GRADE_3, 10—Grade_3.

4.3.4 Design of the Interval Type-2 FS for Classification with Triangular Membership Functions

The following sections specify each of the classifiers that have been performed with type-2 fuzzy systems and triangular membership functions, the structure of the fuzzy system and the parameters used for each of them, as well as the number of fuzzy rules and fuzzy system type.

4.3.4.1   Design of the First Interval Type-2 FS for the Classification of BP Levels with Triangular Membership Functions

The structure of the fuzzy system is illustrated in Fig. 4.36. The numbers marked in the Fig. 4.37 list each of the membership functions for the input systolic and these are: 1—Low, 2—Low_Normal, 3—Normal, 4—High_Normal, 5—High, 6—Very_High, 7—Too_High.

The numbers marked in Fig. 4.38 list each of the MFs for the input diastolic and these are: 1—Low, 2—Low_Normal, 3—Normal, 4—High_Normal, 5—High, 6—Very_High, 7—Too_High.

The numbers marked in Fig. 4.39 list each of the membership functions for the output BP_Levels and these are: 1—Hypotension, 2—Optimal, 3—Normal, 4—High_Normal, 5—ISH_GRADE_1, 6—Grade_1, 7—ISH_GRADE_2, 8—Grade_2, 9—ISH_GRADE_3, 10—Grade_3.

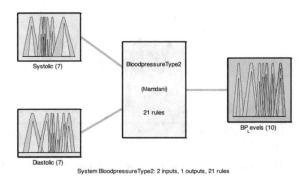

Fig. 4.36 Structure of the interval type-2 FS for classification with triangular membership functions

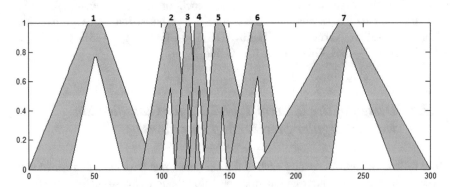

Fig. 4.37 Systolic input for the interval type-2 FS for classification with triangular membership functions

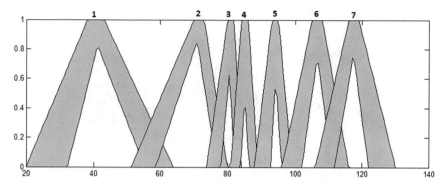

Fig. 4.38 Diastolic input for the interval type-2 FS for classification with triangular membership functions

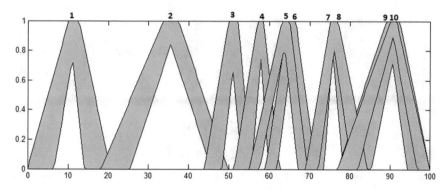

Fig. 4.39 BP_Levels output for the interval type-2 FS for classification with triangular membership functions

Systolic Input																																									
Low						Low_Normal						Normal						High_Normal						High						Very_High						Too_High					
1	2	3	4	5	6	7	8	9	10	11	12	13	14	15	16	17	18	19	20	21	22	23	24	25	26	27	28	29	30	31	32	33	34	35	36	37	38	39	40	41	42

Systolic Input																																									
Low						Low_Normal						Normal						High_Normal						High						Very_High						Too_High					
43	44	45	46	47	48	49	50	51	52	53	54	55	56	57	58	59	60	61	62	63	64	65	66	67	68	69	70	71	72	73	74	75	76	77	78	79	80	81	82	83	84

BP_Levels Output																																									
Hypotension						Optimal						Normal						High_Normal						ISH_Grade_1						Grade_1						ISH_Grade_2					
85	86	87	88	89	90	91	92	93	94	95	96	97	98	99	100	101	102	103	104	105	106	107	108	109	110	111	112	113	114	115	116	117	118	119	120	121	122	123	124	125	126

Grade_2						ISH_Grade_3						Grade_3					
127	128	129	130	131	132	133	134	135	136	137	138	139	140	141	142	143	144

Fig. 4.40 Structure of the chromosome for the interval type-2 fuzzy system for classification with triangular membership functions

Optimization of the Triangular Interval Type-2 Fuzzy Inference System with Genetic Algorithm

The interval type-2 fuzzy system was optimized with a GA, where we have a chromosome to optimize the membership functions (MFS), as shown in Fig. 4.40 and the chromosome has 144 genes, which help to optimize the membership functions. In this case, Genes 1–42 (real numbers) allow to manage the parameters of the systolic input,

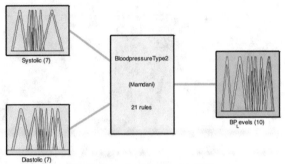

Fig. 4.41 Structure of the optimized interval type-2 fuzzy system for classification with triangular membership functions

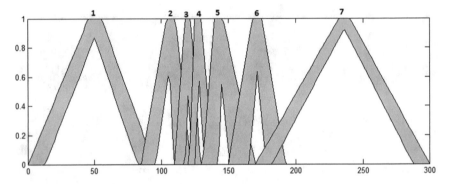

Fig. 4.42 Systolic input for the optimized interval type-2 fuzzy system for classification with triangular membership functions

Genes 43–84 (real numbers) allow to manage the parameters of the diastolic input and Genes 85–144 (real numbers) allow to manage the parameters of the BP_Levels output. The following Fig. 4.40 shows the structure of the chromosome.

Design of the Optimized Interval Type-2 FS for Classification with Triangular Membership Functions

The structure of the optimized interval type-2 fuzzy system is illustrated in Fig. 4.41. The numbers marked in the Fig. 4.42 list each of the membership functions for the input systolic and these are: 1—Low, 2—Low_Normal, 3—Normal, 4—High_Normal, 5—High, 6—Very_High, 7—Too_High.

The numbers marked in the Fig. 4.43 list each of the membership functions for the diastolic input and these are: 1—Low, 2—Low_Normal, 3—Normal, 4—High_Normal, 5—High, 6—Very_High, 7—Too_High.

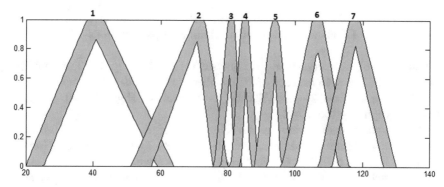

Fig. 4.43 Diastolic input for the optimized interval type-2 fuzzy system for classification with triangular membership functions

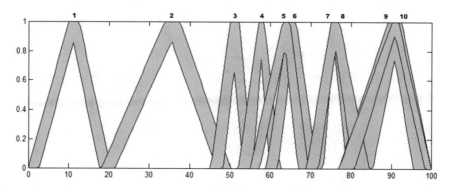

Fig. 4.44 BP_Levels output for the interval type-2 fuzzy system for classification with triangular membership functions

The numbers marked in Fig. 4.44 list each of the membership functions for the output BP_Levels and these are: 1—Hypotension, 2—Optimal, 3—Normal, 4—High_Normal, 5—ISH_GRADE_1, 6—Grade_1, 7—ISH_GRADE_2, 8—Grade_2, 9—ISH_GRADE_3, 10—Grade_3.

4.3.5 Design of the Interval Type-2 FS for Classification with Trapezoidal Membership Functions

The following sections specify each of the classifiers that have been designed with type-2 fuzzy systems and trapezoidal membership functions, the structure of the fuzzy system and the parameters used for each of them, as well as the number of fuzzy rules and fuzzy system type.

4.3.5.1 Design of the First Interval Type-2 FS for the Classification of BP Levels with Trapezoidal Memberships Functions

The structure of the interval type-2 fuzzy system is shown in Fig. 4.45. The numbers marked in the Fig. 4.46 list each of the membership functions for the input systolic and these are: 1—Low, 2—Low_Normal, 3—Normal, 4—High_Normal, 5—High, 6—Very_High, 7—Too_High.

The numbers marked in Fig. 4.47 list each of the membership functions for the input diastolic and these are: 1—Low, 2—Low_Normal, 3—Normal, 4—High_Normal, 5—High, 6—Very_High, 7—Too_High.

The numbers marked in Fig. 4.48 list each of the membership functions for the output BP_Levels and these are: 1—Hypotension, 2—Optimal, 3—Normal, 4—High_Normal, 5—ISH_GRADE_1, 6—Grade_1, 7—ISH_GRADE_2, 8—Grade_2, 9—ISH_GRADE_3, 10—Grade_3.

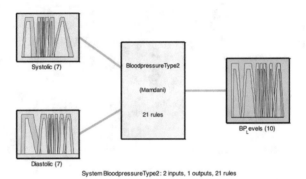

Fig. 4.45 Structure of the interval type-2 FS for classification with trapezoidal MFs

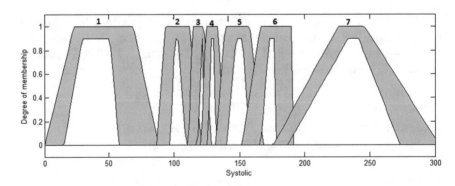

Fig. 4.46 Systolic input for the interval type-2 FS for classification with trapezoidal membership functions (MFs)

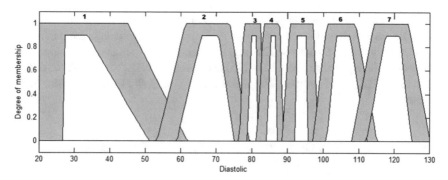

Fig. 4.47 Diastolic input for the interval type-2 FS for classification with trapezoidal MFs

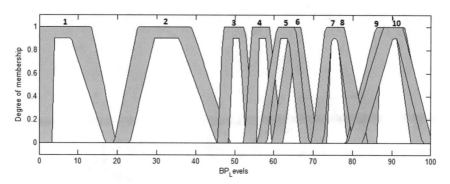

Fig. 4.48 BP_Levels output for the interval type-2 fuzzy system for classification with trapezoidal membership functions

4.3.5.2 Optimization of the Trapezoidal Type-2 Fuzzy Inference System with Genetic Algorithm

The interval type-2 fuzzy system was optimized with a GA, in the GA it is necessary a chromosome to optimize the membership functions (MFS), as shown in Fig. 4.49

Systolic Input																																			
Low									Low_Normal									Normal									High_Normal								
1	2	3	4	5	6	7	8	9	10	11	12	13	14	15	16	17	18	19	20	21	22	23	24	25	26	27	28	29	30	31	32	33	34	35	36
High																		Too_High																	
37	38	39	40	41	42	43	44	45	46	47	48	49	50	51	52	53	54	55	56	57	58	59	60	61	62	63									

Diastolic Input																																			
Low									Low_Normal									Normal									High_Normal								
64	65	66	67	68	69	70	71	72	73	74	75	76	77	78	79	80	81	82	83	84	85	86	87	88	89	90	91	92	93	94	95	96	97	98	99
High									Very_High									Too_High																	
100	101	102	103	104	105	106	107	108	109	110	111	112	113	114	115	116	117	118	119	120	121	122	123	124	125	126									

BP_Levels																																			
Hypotension									Optimal									Normal									High_Normal								
127	128	129	130	131	132	133	134	135	136	137	138	139	140	141	142	143	144	145	146	147	148	149	150	151	152	153	154	155	156	157	158	159	160	161	162
ISH_Grade_1									Grade_1									ISH_Grade_2									Grade_2								
163	164	165	166	167	168	169	170	171	172	173	174	175	176	177	178	179	180	181	182	183	184	185	186	187	188	189	190	191	192	193	194	195	196	197	198
ISH_Grade_3									Grade_3																										
199	200	201	202	203	204	205	206	207	208	209	210	211	212	213	214	215	216																		

Fig. 4.49 Structure of the chromosome for the interval type-2 FS for classification with trapezoidal MFs

and the chromosome has 246 genes and this data help to optimize the membership functions. In this case, Genes 1–63 (real numbers) allow to manage the parameters of the systolic input, Genes 64–126 (real numbers) allow to manage the parameters of the diastolic input and Genes 127–216 (real numbers) allow to manage the parameters of the BP_Levels output. The following Fig. 4.49 shows the structure of the chromosome.

4.3.5.3   Design of the Optimized Interval Type-2 FS for Classification with Trapezoidal MFs

The structure of the optimized interval type-2 fuzzy system is shown in Fig. 4.50. The numbers marked in the Fig. 4.51 list each of the membership functions for the input systolic and these are: 1—Low, 2—Low_Normal, 3—Normal, 4—High_Normal, 5—High, 6—Very_High, 7—Too_High.

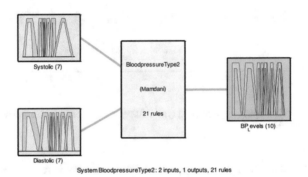

Fig. 4.50 Structure of the optimized interval type-2 FS for classification with trapezoidal MFs

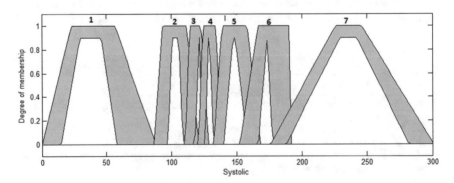

Fig. 4.51 Systolic input for the optimized interval type-2 fuzzy system for classification with trapezoidal membership functions

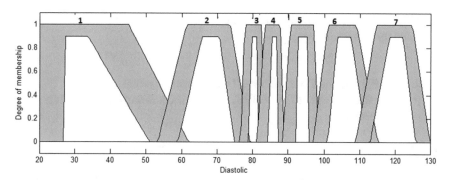

Fig. 4.52 Diastolic input for the optimized interval type-2 fuzzy system for classification with trapezoidal membership functions

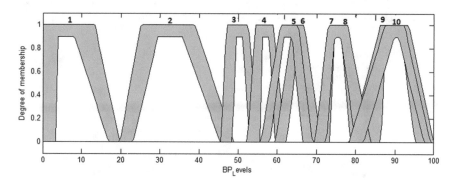

Fig. 4.53 BP_Levels output for the optimized interval type-2 fuzzy system for classification with trapezoidal membership functions

The numbers marked in Fig. 4.52 list each of the membership functions for the input diastolic and these are: 1—Low, 2—Low_Normal, 3—Normal, 4—High_Normal, 5—High, 6—Very_High, 7—Too_High.

The numbers marked in Fig. 4.53 list each of the membership functions for the output BP_Levels and these are: 1—Hypotension, 2—Optimal, 3—Normal, 4—High_Normal, 5—ISH_GRADE_1, 6—Grade_1, 7—ISH_GRADE_2, 8—Grade_2, 9—ISH_GRADE_3, 10—Grade_3.

4.3.6 Design of the Interval Type-2 FS for Classification with Gaussian Membership Functions

The following sections specify each of the classifiers that have been designed with type-2 fuzzy systems and Gaussian membership functions, the structure of the fuzzy system and the parameters used for each of them, as well as the number of fuzzy rules and fuzzy system type.

4.3.6.1 Design of the First Interval Type-2 Fuzzy System for the Classification of BP Levels with Gaussians Memberships Functions

The structure of the interval type-2 fuzzy system with Gaussian MFs is illustrated in Fig. 4.54. The numbers marked in the Fig. 4.55 list each of the membership functions for the input systolic and these are: 1—Low, 2—Low_Normal, 3—Normal, 4—High_Normal, 5—High, 6—Very_High, 7—Too_High.

The numbers marked in Fig. 4.56 list each of the membership functions for the input diastolic and these are: 1—Low, 2—Low_Normal, 3—Normal, 4—High_Normal, 5—High, 6—Very_High, 7—Too_High.

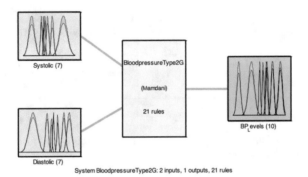

Fig. 4.54 Structure of the interval type-2 FS for classification with Gaussian membership functions (MFs)

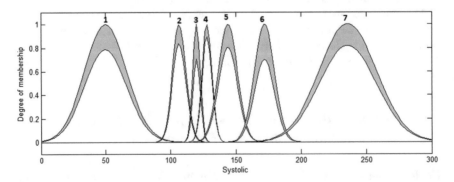

Fig. 4.55 Systolic input for the interval type-2 fuzzy system for classification with Gaussian membership functions

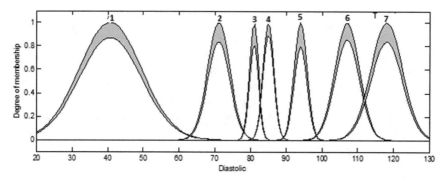

Fig. 4.56 Diastolic input for the interval type-2 fuzzy system for classification with Gaussian membership functions

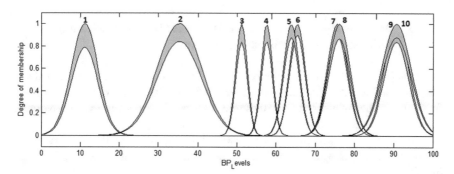

Fig. 4.57 BP_Levels output for the interval type-2 fuzzy system for classification with Gaussians membership functions

The numbers marked in Fig. 4.57 list each of the membership functions for the output BP_Levels and these are: 1—Hypotension, 2—Optimal, 3—Normal, 4—High_Normal, 5—ISH_GRADE_1, 6—Grade_1, 7—ISH_GRADE_2, 8—Grade_2, 9—ISH_GRADE_3, 10—Grade_3.

4.3.6.2 Optimization of the Gaussian Type-2 Fuzzy Inference System with Genetic Algorithm

The interval type-2 fuzzy system was optimized with a GA, in the GA it is necessary a chromosome to optimize the membership functions (MFS), as shown in Fig. 4.58 and the chromosome has 72 genes and this genes help to optimize the membership functions. In this case, Genes 1–21 (real numbers) allow to manage the parameters of the systolic input, Genes 22–42 (real numbers) allow to manage the parameters of the diastolic input and Genes 43–72 (real numbers) allow to manage the parameters of the BP_Levels output. The following Fig. 4.58 shows the structure of the chromosome.

Systolic Input																				
Low			Low_Normal			Normal			High_Normal			High			Very_High			Too_High		
1	2	3	4	5	6	7	8	9	10	11	12	13	14	15	16	17	18	19	20	21
Diastolic Input																				
Low			Low_Normal			Normal			High_Normal			High			Very_High			Too_High		
22	23	24	25	26	27	28	29	30	31	32	33	34	35	36	37	38	39	40	41	42
BP_Levels																				
Hypotension			Low_Normal			Normal			High_Normal			ISH_Grade_1			Grade_1			ISH_Grade_2		
43	44	45	46	47	48	49	50	51	52	53	54	55	56	57	58	59	60	61	62	63
Grade_2			ISH_Grade_3			Grade_3															
64	65	66	67	68	69	70	71	72												

Fig. 4.58 Structure of the chromosome for the interval type-2 FS for classification with Gaussian membership functions

4.3.6.3 Design of the Optimized Interval Type 2 FS for Classification with Gaussians Membership Functions

The structure of the optimized interval type-2 fuzzy system with Gaussian MFs is illustrated in Fig. 4.59. The numbers marked in Fig. 4.60 list each of the membership functions for the input systolic and these are: 1—Low, 2—Low_Normal, 3—Normal, 4—High_Normal, 5—High, 6—Very_High, 7—Too_High.

The numbers marked in Fig. 4.61 list each of the MFs for the input diastolic and these are: 1—Low, 2—Low_Normal, 3—Normal, 4—High_Normal, 5—High, 6—Very_High, 7—Too_High.

The numbers marked in Fig. 4.62 list each of the membership functions for the output BP_Levels and these are: 1—Hypotension, 2—Optimal, 3—Normal, 4—High_Normal, 5—ISH_GRADE_1, 6—Grade_1, 7—ISH_GRADE_2, 8—Grade_2, 9—ISH_GRADE_3, 10—Grade_3.

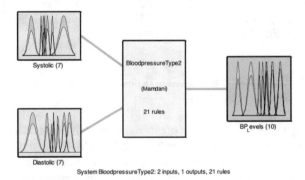

Fig. 4.59 Structure of the optimized interval type-2 FS for classification with Gaussian MFs

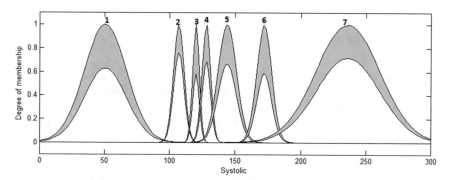

Fig. 4.60 Systolic input for the optimized interval type-2 FS for classification with Gaussian membership functions

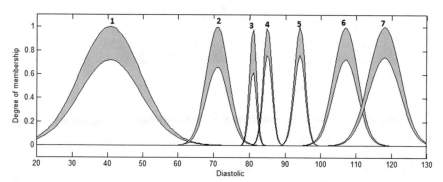

Fig. 4.61 Diastolic input for the optimized interval type-2 fuzzy system for classification with Gaussian membership functions

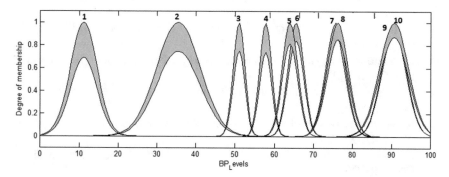

Fig. 4.62 BP_Levels output for the optimized interval type-2 FS for classification with Gaussian membership functions

4.3.7 Fuzzy Rules for the Type-1 and Interval Type-2 FS with the Different Architectures

In previous work we have worked with the optimization of fuzzy rules, for which we already have the optimal number of fuzzy rules based on the optimization made previously, this is why the fuzzy rules are listed below, it should be noted that the same fuzzy rules were used for each architecture of the different type 1 and type 2 fuzzy systems.

1. If (Systolic is Low) and (Diastolic is Low) then (BP_Levels is Hypotension)
2. If (Systolic is Low_Normal) and (Diastolic is Low_Normal) then (BP_Levels is Optimal)
3. If (Systolic is Normal) and (Diastolic is Normal) then (BP_Levels is Normal)
4. If (Systolic is High_Normal) and (Diastolic is High_Normal) then (BP_Levels is High_Normal)
5. If (Systolic is High) and (Diastolic is High) then (BP_Levels is Grade_1)
6. If (Systolic is Very_high) and (Diastolic is Very_High) then (BP_Levels is Grade_2)
7. If (Systolic is too_high) and (Diastolic is Too_High) then (BP_Levels is Grade_3)
8. If (Systolic is Very_high) and (Diastolic is High) then (BP_Levels is Grade_2)
9. If (Systolic is too_high) and (Diastolic is Very_High) then (BP_Levels is Grade_3)
10. If (Systolic is too_high) and (Diastolic is High) then (BP_Levels is Grade_3)
11. If (Systolic is High) and (Diastolic is Very_High) then (BP_Levels is Grade_2)
12. If (Systolic is High) and (Diastolic is Too_High) then (BP_Levels is Grade_3)
13. If (Systolic is Very_high) and (Diastolic is Too_High) then (BP_Levels is Grade_3)
14. If (Systolic is High) and (Diastolic is Normal) then (BP_Levels is ISH_GRADE_1)
15. If (Systolic is High) and (Diastolic is High_Normal) then (BP_Levels is ISH_GRADE_1)
16. If (Systolic is Very_high) and (Diastolic is Normal) then (BP_Levels is ISH_GRADE_2)
17. If (Systolic is Very_high) and (Diastolic is High_Normal) then (BP_Levels is ISH_GRADE_2)
18. If (Systolic is too_high) and (Diastolic is Normal) then (BP_Levels is ISH_GRADE_3)
19. If (Systolic is too_high) and (Diastolic is High_Normal) then (BP_Levels is ISH_GRADE_3)
20. If (Systolic is Normal) or (Diastolic is Normal) then (BP_Levels is Normal)
21. If (Systolic is High_Normal) or (Diastolic is High_Normal) then (BP_Levels is High_Normal)

4.3.8 Knowledge Representation of the Optimized Type-1 and Interval Type-2 Fuzzy Systems

In this part, we show the knowledge representation of the type-1 FS with triangular membership functions. The crisp output is calculated as follows: If the number of fired rules is **r** then the final BP level is:

$$BP = \frac{\sum_{i=1}^{r} BP_i L_i}{\sum_{i=1}^{r} L_i} \tag{4.6}$$

where L_i is the firing level and BP_i is the crisp output of the if-th rule. The triangular curve is a function of a vector, x and depends on three scalar parameters a, b and c, as given by

$$f(x; a, b, c) = \begin{cases} 0 & x \le a \\ \frac{x-a}{b-a}, & a \le x \le b \\ \frac{c-x}{c-b}, & b \le x \le c \\ 0, & c \le x \end{cases} \tag{4.7}$$

4.3.8.1 Input and Output Variables for Triangular Type-1 Fuzzy System

To design a FS, we must determine the input and output linguistic variables, in this FS, we have two inputs and one output.

4.3.8.2 Input Variables

Systolic. This variable has the following membership functions (MFs): Low, Low_normal, Normal, High_normal, High, Very_high, Too_high. These MFs are listed below in Table 4.16.

Diastolic. This variable has the following membership functions (MFS): Low, Low_normal, Normal, High_normal, High, Very_high, Too_high. This MFS are listed below in the Table 4.17.

4.3.8.3 Output Variable

The FS has only one output called **BP_LEVELS**, which refers to the BP level, the output gives a percentage, which refers to the following BP levels: Hypotension,

Table 4.16 Membership functions for the first input variable

$$\mu Low(x) = \begin{bmatrix} 0, & x \le 20.2 \\ \frac{x-20.2}{50.15}, & 20.2 \le x \le 70.35 \\ \frac{94.21-x}{23.86}, & 70.35 \le x \le 94.21 \\ 0, & 94.21 \le x \end{bmatrix}$$

$$\mu Low\_Normal(x) = \begin{bmatrix} 0, & x \le 90.15 \\ \frac{x-90.15}{25.06}, & 90.15 \le x \le 115.21 \\ \frac{122.3-x}{7.09}, & 115 \le x \le 122.3 \\ 0, & 122.3 \le x \end{bmatrix}$$

$$\mu Normal(x) = \begin{bmatrix} 0, & x \le 110.25 \\ \frac{x-110.25}{9.86}, & 110.25 \le x \le 120.11 \\ \frac{129.1-x}{8.99}, & 120.11 \le x \le 129.1 \\ 0, & 129 \le x \end{bmatrix}$$

$$\mu High\_Normal(x) = \begin{bmatrix} 0, & x \le 121.1 \\ \frac{x-121.1}{9.11}, & 121 \le x \le 130.21 \\ \frac{139.2-x}{8.99}, & 130.21 \le x \le 139.2 \\ 0, & 139.2 \le x \end{bmatrix}$$

$$\mu High(x) = \begin{bmatrix} 0, & x \le 131 \\ \frac{x-131}{9.3}, & 131 \le x \le 140.3 \\ \frac{170.17-x}{29.87}, & 140.3 \le x \le 170.17 \\ 0, & 170.17 \le x \end{bmatrix}$$

$$\mu Very\_high(x) = \begin{bmatrix} 0, & x \le 150.23 \\ \frac{x-150.23}{9.77}, & 150.23 \le x \le 160 \\ \frac{192.12-x}{32.12}, & 160 \le x \le 192.12 \\ 0, & 192.12 \le x \end{bmatrix}$$

$$\mu Too\_high(x) = \begin{bmatrix} 0, & x \le 170.35 \\ \frac{x-170.35}{49.78}, & 170.35 \le x \le 220.1 \\ \frac{300.26-x}{80.16}, & 220.1 \le x \le 300.26 \\ 0, & 300.26 \le x \end{bmatrix}$$

Optimal, Normal, High_Normal, Grade 1, Grade 2, Grade 3, Isolated systolic hypertension (ISH) Grade 1, ISH Grade 2 and ISH Grade 3. The linguistic values of the MFs are shown below in Table 4.18.

Table 4.17 Type-1 membership functions for the second input variable

$$\mu Low(x) = \begin{bmatrix} 0, & x \le 20.16 \\ \frac{x-20.16}{20.04}, & 20.16 \le x \le 40.2 \\ \frac{64.1-x}{23.9}, & 40.2 \le x \le 64.1 \\ 0, & 64.1 \le x \end{bmatrix}$$

$$\mu Low\_Normal(x) = \begin{bmatrix} 0, & x \le 60.2 \\ \frac{x-60.2}{9.93}, & 60.2 \le x \le 70.13 \\ \frac{80.22-x}{10.09}, & 70.13 \le x \le 80.22 \\ 0, & 80.22 \le x \end{bmatrix}$$

$$\mu Normal(x) = \begin{bmatrix} 0, & x \le 76.07 \\ \frac{x-76.07}{4.03}, & 76.07 \le x \le 80.1 \\ \frac{84.2-x}{4.1}, & 80.1 \le x \le 84.2 \\ 0, & 84.2 \le x \end{bmatrix}$$

$$\mu High\_Normal(x) = \begin{bmatrix} 0, & x \le 81.2 \\ \frac{x-81.2}{3.8}, & 81.2 \le x \le 85 \\ \frac{89.2-x}{4.2}, & 85 \le x \le 89.2 \\ 0, & 89.2 \le x \end{bmatrix}$$

$$\mu High(x) = \begin{bmatrix} 0, & x \le 88.14 \\ \frac{x-88.14}{1.96}, & 88.14 \le x \le 90.1 \\ \frac{99.1-x}{9}, & 90.1 \le x \le 99.1 \\ 0, & 99.1 \le x \end{bmatrix}$$

$$\mu Very\_high(x) = \begin{bmatrix} 0, & x \le 96.32 \\ \frac{x-96.32}{3.9}, & 96.32 \le x \le 100.22 \\ \frac{115.31-x}{15.09}, & 100.22 \le x \le 115.31 \\ 0, & 115.31 \le x \end{bmatrix}$$

$$\mu Too\_high(x) = \begin{bmatrix} 0, & x \le 107.15 \\ \frac{x-107.15}{2.85}, & 107.15 \le x \le 110 \\ \frac{130.1-x}{20.1}, & 110 \le x \le 130.1 \\ 0, & 130.1 \le x \end{bmatrix}$$

4.3.9 Knowledge Representation of Triangular, Trapezoidal and Gaussian Type-2 Membership Function for Interval Type-2 Fuzzy Systems

To make a fuzzy system, we must first determine the input and output linguistic variables; in this case, we have two inputs and one output.

Table 4.18 Type-1 membership functions for the output variable

$$\mu Hypotension(x) = \begin{cases} 0, & x \leq 0 \\ \frac{x-0}{10.3}, & 0 \leq x \leq 10.3 \\ \frac{20.5-x}{10.2}, & 10.3 \leq x \leq 20.5 \\ 0, & 20.5 \leq x \end{cases}$$

$$\mu Optimal(x) = \begin{cases} 0, & x \leq 20.5 \\ \frac{x-20.5}{19.5}, & 20.5 \leq x \leq 40 \\ \frac{50.3-x}{10.3}, & 40 \leq x \leq 50.3 \\ 0, & 50.3 \leq x50.3 \leq x \end{cases}$$

$$\mu Normal(x) = \begin{cases} 0, & x \leq 49.3 \\ \frac{x-49.3}{3.2}, & 49.3 \leq x \leq 52.5 \\ \frac{54.7-x}{2.2}, & 52.5 \leq x \leq 54.7 \\ 0, & 54.7 \leq x \end{cases}$$

$$\mu High\_Normal(x) = \begin{cases} 0, & x \leq 54.25 \\ \frac{x-54.25}{3.25}, & 54.25 \leq x \leq 57.5 \\ \frac{61.6-x}{4.1}, & 57.5 \leq x \leq 61.6 \\ 0, & 61.6 \leq x \end{cases}$$

$$\mu Grade\_1(x) = \begin{cases} 0, & x \leq 59.1 \\ \frac{x-59.1}{6.4}, & 59.1 \leq x \leq 65.5 \\ \frac{71.5-x}{6}, & 65.5 \leq x \leq 71.5 \\ 0, & 71.5 \leq x \end{cases}$$

$$\mu Grade\_2(x) = \begin{cases} 0, & x \leq 69.4 \\ \frac{x-69.4}{5.6}, & 69.4 \leq x \leq 75 \\ \frac{83.4-x}{8.4}, & 75 \leq x \leq 83.4 \\ 0, & 83.4 \leq x \end{cases}$$

$$\mu Grade\_3(x) = \begin{cases} 0, & x \leq 77.1 \\ \frac{x-77.1}{8.1}, & 77.1 \leq x \leq 85.2 \\ \frac{100-x}{14.8}, & 85.2 \leq x \leq 100 \\ 0, & 100 \leq x \end{cases}$$

$$\mu ISH\_Grade\_1(x) = \begin{cases} 0, & x \leq 55.5 \\ \frac{x-55.5}{7.5}, & 55.5 \leq x \leq 63 \\ \frac{69.4-x}{6.4}, & 63 \leq x \leq 69.4 \\ 0, & 69.4 \leq x \end{cases}$$

(continued)

Table 4.18 (continued)

$$\mu Hypotension(x) = \begin{bmatrix} 0, & x \le 0 \\ \frac{x-0}{10.3}, & 0 \le x \le 10.3 \\ \frac{20.5-x}{10.2}, & 10.3 \le x \le 20.5 \\ 0, & 20.5 \le x \end{bmatrix}$$

$$\mu ISH\_Grade\_2(x) = \begin{bmatrix} 0, & x \le 69.3 \\ \frac{x-69.3}{5.95}, & 69.3 \le x \le 75.25 \\ \frac{83.5-x}{8.25}, & 75.25 \le x \le 83.5 \\ 0, & 83.5 \le x \end{bmatrix}$$

$$\mu ISH\_Grade\_3(x) = \begin{bmatrix} 0, & x \le 77.2 \\ \frac{x-77.2}{8.1}, & 77.2 \le x \le 85.3 \\ \frac{100-x}{14.85}, & 85.3 \le x \le 100 \\ 0, & 100 \le x \end{bmatrix}$$

4.3.9.1 Triangle Interval Type-2 Membership Functions with Uncertainty $a \in [a_1, a_2]$, $b \in [b_1, b_2]$ and $c \in [c_1, c_2]$.

Equation 4.8 represents the triangle Interval Type-2 Membership Functions with Uncertainty.

$$\tilde{\mu}(x) = \left[\underline{\mu}(x), \overline{\mu}(x)\right] = \text{itritype2}(x, [a_1, b_1, c_1, a_2, b_2, c_2]), \text{ where } a_1 < a_2, b_1 < b_2, c_1 < c_2$$

$$\mu_1(x) = \max\left(\min\left(\frac{x-a_1}{b_1-a_1}, \frac{c_1-x}{c_1-b_1}\right), 0\right)$$

$$\mu_2(x) = \max\left(\min\left(\frac{x-a_2}{b_2-a_2}, \frac{c_2-x}{c_2-b_2}\right), 0\right)$$

$$\overline{\mu}(x) = \begin{cases} max(\mu_1(x), \mu_2(x)) & \forall x \notin (b1, b2) \\ 1 & \forall x \in (b1, b2) \end{cases}$$

$$\underline{\mu}(x) = min(\mu_1(x), \mu_2(x)) \tag{4.8}$$

Input Variables for Triangular Type-2 Fuzzy System

Systolic. This variable has the following membership functions (MFS): Low, Low_Normal, Normal, High_Normal, High, Very High and Too_High. This MFS are listed below in the Table 4.19.

   Diastolic. This variable has the following membership functions (MFS): Low, Low_Normal, Normal, High_Normal, High, Very High and Too_High. This MFS are listed below in the Table 4.20.

Table 4.19 Type-1 membership functions for the output variable

Low

$$\mu_1(x) = \max\left(\min\left(\frac{x-0}{45.5}, \frac{82.48-x}{36.98}\right), 0\right)$$

$$\mu_2(x) = \max\left(\min\left(\frac{x-12.8}{42.8}, \frac{94.5-x}{38.9}\right), 0\right)$$

Low_Normal

$$\mu_1(x) = \max\left(\min\left(\frac{x-85.21}{19.09}, \frac{110.3-x}{6}\right), 0\right)$$

$$\mu_2(x) = \max\left(\min\left(\frac{x-96.04}{15.26}, \frac{122.3-x}{11}\right), 0\right)$$

Normal

$$\mu_1(x) = \max\left(\min\left(\frac{x-110.1}{8.2}, \frac{123-x}{4.7}\right), 0\right)$$

$$\mu_2(x) = \max\left(\min\left(\frac{x-118.1}{4.2}, \frac{129.2-x}{6.9}\right), 0\right)$$

High_Normal

$$\mu_1(x) = \max\left(\min\left(\frac{x-121.1}{3.9}, \frac{132.7-x}{7.7}\right), 0\right)$$

$$\mu_2(x) = \max\left(\min\left(\frac{x-125}{5.3}, \frac{139.2-x}{8.9}\right), 0\right)$$

High

$$\mu_1(x) = \max\left(\min\left(\frac{x-131}{9.2}, \frac{151.6-x}{11.4}\right), 0\right)$$

$$\mu_2(x) = \max\left(\min\left(\frac{x-142.6}{3.6}, \frac{170.1-x}{23.9}\right), 0\right)$$

Very_High

$$\mu_1(x) = \max\left(\min\left(\frac{x-150}{18}, \frac{178-x}{10}\right), 0\right)$$

$$\mu_2(x) = \max\left(\min\left(\frac{x-165}{11}, \frac{192-x}{16}\right), 0\right)$$

Too_High

$$\mu_1(x) = \max\left(\min\left(\frac{x-170.5}{62}, \frac{287.8-x}{55.3}\right), 0\right)$$

$$\mu_2(x) = \max\left(\min\left(\frac{x-165}{57.8}, \frac{192-x}{59.5}\right), 0\right)$$

Table 4.20 Type-1 membership functions for the output variable

Low

$$\mu_1(x) = \max\left(\min\left(\frac{x-20}{18.3}, \frac{58.95-x}{20.65}\right), 0\right)$$

$$\mu_2(x) = \max\left(\min\left(\frac{x-25.4}{17.9}, \frac{64.2-x}{20.9}\right), 0\right)$$

Low_Normal

$$\mu_1(x) = \max\left(\min\left(\frac{x-51.4}{18.6}, \frac{75.57-x}{5.57}\right), 0\right)$$

$$\mu_2(x) = \max\left(\min\left(\frac{x-57.3}{15.7}, \frac{80-x}{7}\right), 0\right)$$

Normal

$$\mu_1(x) = \max\left(\min\left(\frac{x-76.1}{4.2}, \frac{82.5-x}{2.2}\right), 0\right)$$

$$\mu_2(x) = \max\left(\min\left(\frac{x-78.3}{3.7}, \frac{84.2-x}{2.2}\right), 0\right)$$

High_Normal

$$\mu_1(x) = \max\left(\min\left(\frac{x-81}{3.5}, \frac{87.3-x}{2.8}\right), 0\right)$$

$$\mu_2(x) = \max\left(\min\left(\frac{x-84}{2.2}, \frac{89-x}{2.8}\right), 0\right)$$

High

$$\mu_1(x) = \max\left(\min\left(\frac{x-88.2}{5.1}, \frac{96-x}{2.7}\right), 0\right)$$

$$\mu_2(x) = \max\left(\min\left(\frac{x-92}{3.5}, \frac{99-x}{3.5}\right), 0\right)$$

Very_High

$$\mu_1(x) = \max\left(\min\left(\frac{x-96.1}{9.2}, \frac{113-x}{7.7}\right), 0\right)$$

$$\mu_2(x) = \max\left(\min\left(\frac{x-100.4}{8}, \frac{116-x}{7.6}\right), 0\right)$$

Too_High

$$\mu_1(x) = \max\left(\min\left(\frac{x-107}{9.5}, \frac{126.2-x}{9.7}\right), 0\right)$$

$$\mu_2(x) = \max\left(\min\left(\frac{x-111.3}{8.2}, \frac{130-x}{10.5}\right), 0\right)$$

Output Variables for Triangular Type-2 Fuzzy System

The FS has only one output called BP_LEVELS, which refers to the Blood pressure (BP) level, the output gives a percentage, which refers to the following BP levels: Hypotension, Optimal, Normal, High_Normal, Grade_1, ISH_Grade1, Grade_2, ISH_Grade2, Grade_3, ISH_Grade3. The linguistic value of MFs is shown below in the Table 4.21.

4.3.9.2 Trapezoidal Interval Type-2 Membership Functions with Uncertain $a \in [a_1, a_2], b \in [b_1, b_2], c \in [c_1, c_2]$ and $d \in [d_1, d_2]$.

Equation 4.9 represents trapezoidal interval type-2 membership functions with uncertain.

$$\tilde{\mu}(x) = \left[\underline{\mu}(x), \overline{\mu}(x)\right] = \text{itrapatype2}(x, [a_1, b_1, c_1, d_1, a_2, b_2, c_2, d_2, \alpha])$$

where $a_1 < a_2, b_1 < b_2, c_1 < c_2, d_1 < d_2$

$$\mu_1(x) = \max\left(\min\left(\frac{x - a_1}{b_1 - a_1}, 1, \frac{d_1 - x}{d_1 - c_1}\right), 0\right)$$

$$\mu_2(x) = \max\left(\min\left(\frac{x - a_2}{b_2 - a_2}, 1, \frac{d_2 - x}{d_2 - c_2}\right), 0\right)$$

$$\overline{\mu}(x) = \begin{cases} max(\mu_1(x), \mu_2(x)) & \forall x \notin (b1, c2) \\ 1 & \forall x \in (b1, c2) \end{cases}$$

$$\underline{\mu}(x) = min(\alpha, min(\mu_1(x), \mu_2(x)))$$

$$(4.9)$$

4.3.9.3 Interval Type-2 Gaussian Membership Function

Equation 4.10 represents Interval Type-2 Gaussian Membership Function.

$$\tilde{\mu}(x) = \left[\underline{\mu}(x), \overline{\mu}(x)\right] = \text{igaussatype2}(x, [\sigma, m, \alpha])$$

$$\underline{\mu}(x) = \alpha \exp\left[-\frac{1}{2}\left(\frac{x - m}{\sigma}\right)^2\right] \quad \text{Where } 0 < \alpha < 1$$

$$\overline{\mu}(x) = \exp\left[-\frac{1}{2}\left(\frac{x - m}{\sigma}\right)^2\right]$$

$$(4.10)$$

Table 4.21 Type-1 membership functions for the output variable

Hypotension

$$\mu_1(x) = \max\left(\min\left(\frac{x-0}{10}, \frac{17.7-x}{7.7}\right), 0\right)$$

$$\mu_2(x) = \max\left(\min\left(\frac{x-2.72}{9.77}, \frac{20.5-x}{8.01}\right), 0\right)$$

Optimal

$$\mu_1(x) = \max\left(\min\left(\frac{x-18.2}{15.5}, \frac{46.7-x}{13}\right), 0\right)$$

$$\mu_2(x) = \max\left(\min\left(\frac{x-21.63}{15.87}, \frac{50.2-x}{12.7}\right), 0\right)$$

Normal

$$\mu_1(x) = \max\left(\min\left(\frac{x-45.27}{4.63}, \frac{53.4-x}{3.5}\right), 0\right)$$

$$\mu_2(x) = \max\left(\min\left(\frac{x-48.3}{4}, \frac{55.5-x}{3.2}\right), 0\right)$$

High_Normal

$$\mu_1(x) = \max\left(\min\left(\frac{x-52.16}{4.84}, \frac{60-x}{3}\right), 0\right)$$

$$\mu_2(x) = \max\left(\min\left(\frac{x-55.6}{2.9}, \frac{62.5-x}{4}\right), 0\right)$$

Grade_1

$$\mu_1(x) = \max\left(\min\left(\frac{x-59.1}{5.3}, \frac{68.4-x}{4}\right), 0\right)$$

$$\mu_2(x) = \max\left(\min\left(\frac{x-61.9}{4.3}, \frac{71.6-x}{5.2}\right), 0\right)$$

ISH_Grade_1

$$\mu_1(x) = \max\left(\min\left(\frac{x-55}{8}, \frac{67-x}{4}\right), 0\right)$$

$$\mu_2(x) = \max\left(\min\left(\frac{x-57.5}{7.4}, \frac{69.5-x}{4.6}\right), 0\right)$$

Grade_2

$$\mu_1(x) = \max\left(\min\left(\frac{x-69}{6}, \frac{79-x}{4}\right), 0\right)$$

$$\mu_2(x) = \max\left(\min\left(\frac{x-73.23}{2.77}, \frac{84.5-x}{8.5}\right), 0\right)$$

(continued)

Table 4.21 (continued)

Hypotension

$$\mu_1(x) = \max\left(\min\left(\frac{x-0}{10}, \frac{17.7-x}{7.7}\right), 0\right)$$

$$\mu_2(x) = \max\left(\min\left(\frac{x-2.72}{9.77}, \frac{20.5-x}{8.01}\right), 0\right)$$

ISH_Grade_2

$$\mu_1(x) = \max\left(\min\left(\frac{x-69}{6}, \frac{80-x}{5.6}\right), 0\right)$$

$$\mu_2(x) = \max\left(\min\left(\frac{x-72.1}{4.9}, \frac{84.5-x}{7.5}\right), 0\right)$$

Grade_3

$$\mu_1(x) = \max\left(\min\left(\frac{x-77}{12}, \frac{95.5-x}{6.5}\right), 0\right)$$

$$\mu_2(x) = \max\left(\min\left(\frac{x-83}{9.4}, \frac{100-x}{7.6}\right), 0\right)$$

ISH_Grade_3

$$\mu_1(x) = \max\left(\min\left(\frac{x-77}{13}, \frac{98.4-x}{8.4}\right), 0\right)$$

$$\mu_2(x) = \max\left(\min\left(\frac{x-80.94}{10.66}, \frac{100-x}{8.4}\right), 0\right)$$

4.3.10 Results of This Work

The results obtained are based on the optimization of the membership functions with the genetic algorithm, in Table 4.22 are shown the best architectures of the genetic algorithm.

The parameters used in the algorithm are generation: 100, population: 100, selection method: roulette wheel, mutation rate: 0.06 crossing rate: 0.5. These are the parameters used, since in previous tests, a good error was obtained using these parameters.

The classification error is based in the fitness function as shown Eq. (4.12), the thinking was to limit the error order and with this realizing that the base classifier

Table 4.22 Parameters that were tested before choosing the optimal parameters for the GA

Genetic algorithm	Generation	Population	Selection method	Mutation rate	Crossing rate
GA 1	**100**	**100**	**Roulette wheel**	**0.06**	**0.5**
GA 2	100	100	Roulette wheel	0.04	0.6
GA 3	100	100	Roulette wheel	0.06	0.7

is arranging effectively, the best way to know if the classifier is working correctly is by guiding the 2.1, which shows the official blood pressure levels. Table 4.22 demonstrates the distinctive parameters utilized as a part of the GA.

The next Tables 4.23 and 4.24 show the results obtained for 30 patients randomly selected and based on these results we obtain the classification accuracy (CA) rate and classification error (CE) rate, for which we use the following equations:

The CA Rate is calculated as follows:

$$CA = \frac{N_c}{N_t} \tag{4.11}$$

where N_c is the Number of Training Instances Correctly Classified and N_t is the Number of Training instances.

The CE is calculated as follows:

$$CE = \frac{N_e}{N_t} \tag{4.12}$$

where N_e is the Number of Training Instances Incorrectly Classified and N_t is the Number of Training instances. The columns shaded with yellow, are the incorrect classifications of each classifier.

Below are the experiments done to obtain the best fuzzy classifier, it should be noted that they were tested with 30 patients and this was the result which is shown in Table 4.23 for type-1 fuzzy systems and the results obtained for interval type-2 fuzzy systems is shown in Table 4.24.

Based on the experiments performed with the different architectures, we were able to compare each one of the results and reach the conclusion that the best architecture is the one that is composed of type-2 triangular membership functions, with 21 fuzzy rules and Mamdani type. It is worth mentioning that all architectures improved when using type-2 but the best one is the type-2 fuzzy system with triangular membership functions.

The results for the classification of the experiments with 45 systolic and diastolic samples from the 24 h monitoring in 30 patients using type-1 fuzzy systems with triangular membership functions is shown in Table 4.25 with an average 98% and the standard Deviation of 2.36, trapezoidal membership functions is: average 91.92% and the standard Deviation of 7.16 and finally Gaussian membership functions is 91.92% and the standard Deviation of 5.91, then the best one is the triangular type-1 fuzzy system with average of 98% and the standard Deviation is 2.36. Below we also show the type-2 fuzzy systems with triangular membership functions results is shown in Table 4.26 with an average of 99.40% and the standard Deviation of 0.99, trapezoidal membership functions is: average 94.29% and the standard Deviation of 4.16 and finally Gaussian membership functions is 93.63% and the standard Deviation of 4.59. The conclusion is that the best one is the type-2 fuzzy system with triangular membership function with the average of 99.40% and the standard Deviation is 0.99.

Table 4.23 Experiments of the 30 patients with triangular, trapezoidal and Gaussian MFs with type-1 fuzzy system

Experiments with 45 systolic and diastolic samples from the 24 h monitoring

Patients	Correct classification percentage of the optimized classifier T1 (triangular)	Correct classification percentage of the optimized classifier T1 (trapezoidal)	Correct classification percentage of the optimized classifier T1 (Gaussians)
1	93.33	97.78	93.33
2	100	93.33	93.33
3	93.33	97.78	93.33
4	100	97.78	80
5	100	93.33	95.56
6	97.78	100	77.78
7	93.33	93.33	95.56
8	97.78	93.33	95.56
9	97.78	97.78	80
10	100	77.78	84.44
11	100	93.33	100
12	97.78	93.33	95.56
13	100	97.78	93.33
14	97.78	93.33	93.33
15	100	97.78	100
16	97.78	100	95.56
17	100	93.33	84.44
18	97.78	93.33	84.44
19	100	77.78	93.33
20	93.33	97.78	93.33
21	100	86.66	95.56
22	100	93.33	93.33
23	97.78	86.66	95.56
24	97.78	93.33	86.66
25	100	77.78	95.56
26	97.78	77.78	95.56
27	100	97.78	95.56
28	97.78	93.33	86.66
29	93.33	77.78	95.56
30	97.78	93.33	95.56
Average	98	91.92	91.92

Table 4.24 Experiments of the 30 patients by the optimized type-2(T2) fuzzy inference system with triangular, trapezoidal and Gaussian MFs based in an expert

Experiments with 45 systolic and diastolic samples from the 24 h monitoring			
Patients	Correct classification percentage of the optimized classifier T2 (triangular)	Correct classification percentage of the optimized classifier T2 (trapezoidal)	Correct classification percentage of the optimized classifier T2 (Gaussians)
1	100	97.78	93.33
2	100	93.33	95.56
3	100	97.78	93.33
4	100	97.78	95.56
5	100	93.33	95.56
6	100	100	77.78
7	97.78	93.33	95.56
8	100	93.33	95.56
9	100	97.78	95.56
10	100	93.33	84.44
11	100	97.78	100
12	97.78	93.33	95.56
13	100	97.78	93.33
14	97.78	93.33	93.33
15	100	97.78	100
16	100	100	95.56
17	100	93.33	84.44
18	100	93.33	93.33
19	100	97.78	93.33
20	100	97.78	93.33
21	100	86.66	95.56
22	100	93.33	93.33
23	100	86.66	95.56
24	97.78	93.33	86.66
25	100	97.78	95.56
26	97.78	93.33	95.56
27	100	86.66	95.56
28	97.78	86.66	95.56
29	97.78	86.66	95.56
30	97.78	97.78	95.56
Average	99.40	94.29	93.63

Table 4.25 Results for the experiments with 45 systolic and diastolic samples from the 24 h monitoring in 30 patients using type-1 FS with triangular, trapezoidal and Gaussian MFs

Results for type-1 fuzzy systems			
	T1 classifier (triangular MF)	T1 classifier (trapezoidal MF)	T1 classifier (Gaussians MF)
Average (%)	98	91.92	91.92
Variance	5.57	51.25	34.92
Standard deviation	2.36	7.16	5.91

Table 4.26 Results for the classification of the 30 experiments in the type-2 FS with triangular, trapezoidal and Gaussian MFs

Results for type-2 fuzzy system			
	T2 classifier (triangular MF)	T2 classifier (trapezoidal MF)	T2 classifier (Gaussians MF)
Average (%)	99.40	94.29	93.63
Variance	0.99	17.29	21.04
Standard deviation	0.99	4.16	4.59

It is also important to highlight the results obtained in the architectures with trapezoidal and Gaussian membership functions, since their standard deviation of type-2 is lower than that of type-1, this is why we conclude that the trapezoidal and Gaussian architectures in type-2 are better than Trapezoidal and Gaussian architectures in Type-1.

4.3.11   Statistical Test

To validate in the best way the proposed method, it was decided to use the statistical z-test, which is given by Eq. 4.13, and the parameters used for the test are shown in Table 4.27.

Table 4.27 Values for the statistical z-test

Parameter	Value
Level of confidence	95%
Alpha	0.05
H_a	$\mu 1 > \mu 2$ (Claim)
H_o	$\mu_1 \leq \mu_2$
Critical value	1.645

$$Z = \frac{(\overline{X}_1 - \overline{X}_2) - (\mu_1 - \mu_2)}{\sigma_{\overline{X}_1 - \overline{X}_2}} \tag{4.13}$$

4.3.11.1 Statistical Test for Type-1 Fuzzy Systems Classifiers

The following statistical test was done to compare the results of type-1 fuzzy classifiers as shown in the Table 4.28.

The alternative hypothesis indicates that the Classifier T1 with triangular membership function is greater than the classifier T1 with trapezoidal membership functions and the null hypothesis indicates otherwise, based on the information provided, the significance level is $\alpha = 0.05$, and the critical value for a right-tailed test is $z_c = 1.64$. Since it is observed that $z = 4.414 > z_c = 1.64$, it is then concluded that the null hypothesis is rejected. Therefore, the claim of the alternative hypothesis is accepted, mentioning that the T1 classifier with triangular membership functions is greater than the T1 classifier with trapezoidal membership functions as shows in Table 4.28. The "S" means that significant evidence was found and "N.S" refers to the fact that no significant evidence was found.

The alternative hypothesis indicates that the Classifier T1 with triangular membership function is greater than the classifier T1 with Gaussians membership functions and the null hypothesis indicates otherwise, based on the information provided, the significance level is $\alpha = 0.05$, and the critical value for a right-tailed test is $z_c = 1.64$. Since it is observed that $z = 5.228 > z_c = 1.64$, it is then concluded that the null hypothesis is rejected. Therefore, the claim of the alternative hypothesis is accepted, mentioning that the T1 classifier with triangular membership functions is greater than the T1 classifier with Gaussian membership functions as shows in Table 4.29.

In conclusion, we can affirm that the fuzzy T1 classifier with triangular membership functions is better than classifiers with other architectures. The statistical test of the best architecture is then carried out with the fuzzy classifier type-1 and the fuzzy classifier type-2 for the classification of blood pressure levels.

Table 4.28 Results of the statistical test of type-1 fuzzy classifiers, triangular versus trapezoidal

T1-classifier with triangular MFs			
Classifier	Zc	Z_Value	Evidence
T1-classifier with trapezoidal MFs	1.64	4.414	S

Table 4.29 Results of statistical test of type-1 fuzzy classifiers, triangular versus Gaussian

T1-classifier with triangular MFs			
Classifier	Zc	Z_Value	Evidence
T1-classifier with Gaussian MFs	1.64	5.228	S

Table 4.30 Results of statistical test of type-2 versus type-1 fuzzy classifiers

T2-classifier with triangular MFs is greater			
Classifier	Zc	Z_Value	Evidence
T1-classifier with triangular MFs	1.64	3.01	S

4.3.11.2   Statistical Test for Type-2 Fuzzy Systems Classifiers

The following statistical test is to compare the best type-1 fuzzy system classifier, in this case is the classifier with triangular MFs and the best type-2 fuzzy system classifier is the classifier with triangular MFs as shown in Table 4.30.

The alternative hypothesis indicates that the Classifier T2 with triangular membership function is greater than the classifier T1 with triangular membership functions and the null hypothesis indicates otherwise, based on the information provided, the significance level is $\alpha = 0.05$, and the critical value for a right-tailed test is $z_c = 1.64$. Since it is observed that $z = 3.01 > z_c = 1.64$, it is then concluded that the null hypothesis is rejected. Therefore, the claim of the alternative hypothesis is accepted, mentioning that the T2 classifier with triangular membership functions is greater than the T1 classifier with triangular membership functions. The "S" means that significant evidence was found and "N.S" refers to the fact that no significant evidence was found.

The results indicated in Table 4.30 demonstrate that the fuzzy type-2 classifier with triangular membership functions is better than the other classifiers analyzed; therefore, there is sufficient significant evidence to reject the null hypothesis. When the z-test with a significance level of 0.05 is applied, the alternative hypothesis indicates that the type-2 classifier with triangular membership functions is greater than the type-1 classifier with triangular membership functions.

4.3.12   Discussion

This work is focused on analyzing each of the possible architectures of type-1 and type-2 fuzzy systems, in order to obtain the best classifier with the least possible error at the moment of making the blood pressure classification.

In the work entitled design of an optimized fuzzy classifier for the diagnosis of blood pressure with a new computational method for expert rule optimization, the design of a type-1 fuzzy classifier was carried out with the optimization of triangular membership functions and the appropriate fuzzy rules based on the knowledge of an expert in cardiology. The optimization was done with genetic algorithms, in this work the chromosome structure is shown for the optimization and thus finding the appropriate parameters. It is worth mentioning that only work was done with type-1 fuzzy systems and it was limited to testing with triangular functions, since the main objective was to obtain the appropriate number of fuzzy rules and thus avoid possible unnecessary rules for the classification of blood pressure.

In the current work, once the optimal number of fuzzy rules was achieved, we decided to design type-1 and type-2 fuzzy classifiers using triangular, trapezoidal and Gaussian membership functions in order to compare the different architectures with which the experiments were carried out. These are shown in the results section and all this in order to obtain the architecture with the lowest classification error rate.

It is important to emphasize that the use of type-2 fuzzy systems can help to improve the results, compared to previous works of the authors, that is why the contribution of this work was to find a better classification architecture based on interval type-2 fuzzy systems, since the management of uncertainty in their membership functions helped to give a more adequate classification.

At present time, there are some works in the literature, which have done research in medicine focused on cancer, diabetes, nutrition, cardiovascular diseases among others, all these using other intelligent techniques [5–8].

4.4 Blood Pressure Load

The constant check-ups help the patient to know if he/she has this silent disease or not, that is why today it is common to use 24-h monitoring, which provides the information of all day blood pressures loads samples, which experts in cardiology have given an important emphasis to this measure, which can help to diagnose this disease more accurately than the levels that are commonly used to give a final result. It is important to emphasize that the use of BP levels is an important risk factor that helps to determine this cardiovascular problem, however some studies done by experts in the subject have led to analyze the blood pressure load.

The blood pressure load is the percentage above the established ranges which indicate a range for the day and another range for the night. The ranges set for day time blood pressure load is $\geq$ 135/85 mmHg and the range set for night time blood pressure load is $\geq$ 120/70 mmHg [9, 10].

Some previous studies have mentioned the relationship that may exist between the blood pressure load and determination of the damage that an organ can have, such as the left ventricular, among others [11, 12]. Most studies have not given much explanation regarding the correlation that may exist between the blood pressure level and the blood pressure load. This is why in this study we chose to use some intelligent computing techniques to determine an accurate diagnosis. First, this work focuses on the development and implementation of a fuzzy systems, which provides a classification of the BP levels and BP loads. In the Fig. 4.63 show the Specific Neuro Fuzzy Hybrid Model for the BP load.

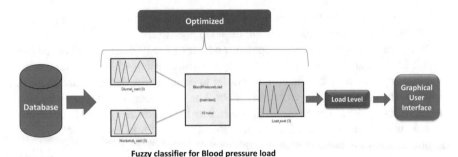

Fig. 4.63 Specific neuro fuzzy hybrid model for blood pressure load

Table 4.31 Blood pressure load categories

Blood pressure load categories	Blood pressure load ranges
Normal load	<20
Intermediate load	20–40
High load	>40

4.4.1 Blood Pressure Load

The blood pressure load is one of the measures that today is incorporated into all those 24 h monitoring devices in order to help the expert to give a better diagnosis, which is why the BP load It has different classification categories. These categories are shown in Table 4.31.

4.4.2 Examples of a Monitoring Record with Blood Pressure Load

The Blood pressure (BP) load is obtained from the number of readings over the range established for day and night. In Table 4.32 we can find an example of a summary of a patient with 35 readings during the day, of which 3 readings are indicated as BP load, and during the night it has 18 readings, of which 15 are taken as BP load, it is therefore, fuzzy system takes these percentages as inputs and based on the highest percentage given as input, is the level of BP load that is diagnosed (Table 4.33).

Table 4.32   Example of a blood pressure load record

Date	Time	Systolic	Diastolic	Pulse
2/2/2016	17:41	135	78	68
2/2/2016	18:01	131	75	63
2/2/2016	18:21	130	70	68
2/2/2016	18:41	117	68	68
2/2/2016	19:01	133	65	66
2/2/2016	19:21	122	77	65
2/2/2016	19:41	117	73	70
2/2/2016	20:07	107	66	70
2/2/2016	20:21	112	62	66
2/2/2016	20:53	122	67	64
2/2/2016	21:01	116	73	63
2/2/2016	21:22	106	71	64
2/2/2016	22:01	103	62	64
2/2/2016	22:21	101	65	68
2/2/2016	23:01	115	62	63
3/2/2016	0:01	114	69	59
3/2/2016	1:01	114	77	62
3/2/2016	1:31	109	66	58
3/2/2016	2:01	116	59	58
3/2/2016	2:31	119	65	55
3/2/2016	3:04	112	64	57
3/2/2016	3:31	98	52	63
3/2/2016	4:04	108	64	55
3/2/2016	4:31	112	61	55
3/2/2016	5:05	116	66	59
3/2/2016	5:31	116	70	55
3/2/2016	6:31	121	70	52
3/2/2016	7:01	112	70	58
3/2/2016	7:21	115	69	58
3/2/2016	7:41	117`	68	55
3/2/2016	8:01	110	72	61
3/2/2016	9:01	129	75	56
3/2/2016	9:21	108	74	55
3/2/2016	9:41	115	74	59
3/2/2016	10:01	134	74	64
3/2/2016	10:21	115	67	64

(continued)

Table 4.32 (continued)

Date	Time	Systolic	Diastolic	Pulse
3/2/2016	10:41	125	69	65
3/2/2016	11:01	115	69	68
3/2/2016	11:21	122	64	66
3/2/2016	11:41	121	66	64
3/2/2016	12:01	114	74	64
3/2/2016	12:41	127	68	65
3/2/2016	13:01	112	76	66
3/2/2016	13:21	119	66	67
3/2/2016	13:41	123	69	64
3/2/2016	14:01	122	69	67
3/2/2016	14:21	129	60	65
3/2/2016	14:41	126	73	65
3/2/2016	15:01	132	75	62
3/2/2016	15:41	120	65	68
3/2/2016	16:22	112	73	80

Table 4.33 Example of a summary of blood pressure load record

Day and night period	Time	Interval	Readings	Readings with BP_load	Awake	Asleep
Day period	07 ~ 22	20 min	38	1	2.60%	
Night period	22 ~ 07	30 min	13	3		23.07%
Day BP load			(% of day readings ≥ 135/85 mmHg)			
Night BP load			(% of night readings ≥ 120/70 mmHg)			

4.4.3   Optimization of Type-1 and Type-2 Fuzzy System for the Classification of Blood Pressure Load

The structure of the type-1 fuzzy system as shown in Fig. 4.64 consists of the following inputs: Diurnal_Load and Nocturnal_Load and an output called Level_Load. The Diurnal_Load input has the following membership functions (MFs): Normal_Diurnal_Load, Intermediate_Diurnal_Load and High_Diurnal_Load. The Nocturnal_Load input has the following membership functions: Normal_Nocturnal_Load, Intermediate_Nocturnal_Load and High_Nocturnal_Load. For the output Level_Load has the following MFs: Normal,

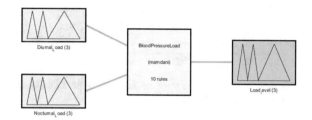

Fig. 4.64 Structure of the type-1 fuzzy system for classification of blood pressure load with Triangular membership functions

Intermediate and High. The fuzzy system is of Mamdani type and has ten fuzzy rules [13, 14].

In the Genetic algorithm, it is necessary to define a chromosome to optimize, in this case the (Membership Functions (MFs) as shown in Fig. 4.65 and the chromosome has 27 genes that help to optimize the MFs. In this case, Genes 1–9 allow to manage the parameters of the Diurnal_Load input, Genes 10–18 allow to manage the parameters of the Nocturnal_Load input and Genes 19–27 allow to manage the parameters of the Load_Level output. The following Fig. 4.65 shows the structure of the chromosome [15–18]:

The structure of the interval type-2 fuzzy system is illustrated in Fig. 4.66 and it has the following inputs: Diurnal_Load and Nocturnal_Load and an

Diurnal_Load								
Normal_Diurnal_Load			Intermediate_Diurnal_Load			High_Diurnal_Load		
1	2	3	4	5	6	7	8	9
Nocturnal_Load								
Normal_Nocturnal_Load			Intermediate_Nocturnal_Load			High_Nocturnal_Load		
10	11	12	13	14	15	16	17	18
Load_Level								
Normal			Intermediate			High		
19	20	21	22	23	24	25	26	27

Fig. 4.65 Structure of the chromosome of the type-1 for classification of blood pressure load with Triangular membership functions

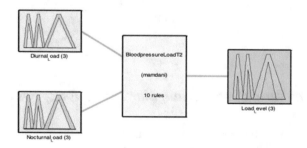

Fig. 4.66 Structure of the interval type-2 fuzzy system for the classification of blood pressure load

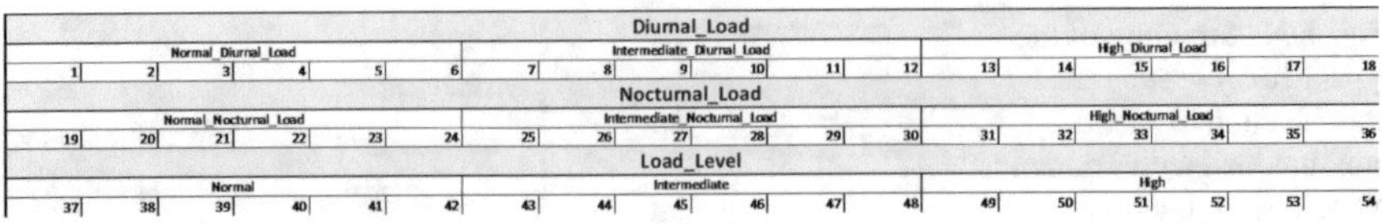

Diurnal_Load																	
Normal_Diurnal_Load						Intermediate_Diurnal_Load								High_Diurnal_Load			
1	2	3	4	5	6	7	8	9	10	11	12	13	14	15	16	17	18

Nocturnal_Load																	
Normal_Nocturnal_Load						Intermediate_Nocturnal_Load								High_Nocturnal_Load			
19	20	21	22	23	24	25	26	27	28	29	30	31	32	33	34	35	36

Load_Level																	
Normal						Intermediate								High			
37	38	39	40	41	42	43	44	45	46	47	48	49	50	51	52	53	54

Fig. 4.67 Structure of the chromosome for the type-2 for classification of blood pressure load with Triangular membership functions

output called Level_Load. The Diurnal_Load input has the following membership functions (MFs): Normal_Diurnal_Load, Intermediate_Diurnal_Load and High_Diurnal_Load. The Nocturnal_Load input has the following membership functions: Normal_Nocturnal_Load, Intermediate_Nocturnal_Load and High_Nocturnal_Load. For the output Level_Load has the following MFs: Normal, Intermediate and High. The fuzzy system is Mamdani type and has ten fuzzy rules.

The chromosome has 54 genes and this help to optimize the MFs, Genes 1–18 allow to manage the parameters of the Diurnal_Load input, Genes 19–36 allow to manage the parameters of the Nocturnal_Load input and Genes 37–54 allow to manage the parameters of the Load_Level output. The following Fig. 4.67 shows the structure of the chromosome of the type-2 fuzzy system for classification of blood pressure load with Triangular membership functions.

4.4.4   Knowledge Representation of the Optimized Type-1 and Interval Type-2 Fuzzy Systems with Triangular Memberships Function

The triangular curve is a function of a vector, x, and depends on three scalar parameters a, b, and c, as given by

$$f(x; a, b, c) = \begin{cases} 0, & x \leq a \\ \frac{x-a}{b-a}, & a \leq x \leq b \\ \frac{c-x}{c-b}, & b \leq x \leq c \\ 0, & c \leq x \end{cases} \tag{4.14}$$

4.4.5   Input Variables for Triangular Type-1 Fuzzy System

1. **Diurnal_Load.** This variable has the following membership functions (MFS): Normal_Diurnal_Load, Intermediate_Diurnal_Load and High_Diurnal_Load. This MFS are listed below in the Table 4.34.

Table 4.34 Type-1 membership functions for the first input variable

$$\mu Normal\_Diurnal\_Load(x) = \begin{cases} 0, & x \le 0 \\ \frac{x-10}{10}, & 0 \le x \le 10 \\ \frac{21.5-x}{11.5}, & 10 \le x \le 21.5 \\ 0, & 21.5 \le x \end{cases}$$

$$\mu Intermediate\_Diurnal\_Load(x) = \begin{cases} 0, & x \le 19.5 \\ \frac{x-19.5}{10.5}, & 19.5 \le x \le 30 \\ \frac{41.5-x}{11.5}, & 30 \le x \le 41.5 \\ 0, & 41.5 \le x \end{cases}$$

$$\mu High\_Diurnal\_Load(x) = \begin{cases} 0, & x \le 39.5 \\ \frac{x-39.5}{30.5}, & 39.5 \le x \le 70 \\ \frac{100-x}{30}, & 70 \le x \le 100 \\ 0, & 100 \le x \end{cases}$$

2. **Nocturnal_Load.** This variable has the following membership functions (MFS): Normal_Nocturnal_Load, Intermediate_Nocturnal_Load and High_Nocturnal_Load. This MFS are listed in Table 4.35.

Output Variables for Triangular Type-1 Fuzzy System:

1. The FS has only one output called **LOAD_LEVEL**, which refers to the Blood pressure (BP) load level, the output gives a percentage, which refers to the following BP Load levels: Normal, Intermediate and High. The linguistic value of MFs is shown in Table 4.36.

Table 4.35 Type-1 Membership functions for the second input variable

$$\mu Normal\_Nocturnal\_Load(x) = \begin{cases} 0, & x \le 0 \\ \frac{x-10}{10}, & 0 \le x \le 10 \\ \frac{21.5-x}{11.5}, & 10 \le x \le 21.5 \\ 0, & 21.5 \le x \end{cases}$$

$$\mu Intermediate\_Nocturnal\_Load(x) = \begin{cases} 0, & x \le 19.5 \\ \frac{x-19.5}{10.5}, & 19.5 \le x \le 30 \\ \frac{41.5-x}{11.5}, & 30 \le x \le 41.5 \\ 0, & 41.5 \le x \end{cases}$$

$$\mu High\_Nocturnal\_Load(x) = \begin{cases} 0, & x \le 39.5 \\ \frac{x-39.5}{30.5}, & 39.5 \le x \le 70 \\ \frac{100-x}{30}, & 70 \le x \le 100 \\ 0, & 100 \le x \end{cases}$$

Table 4.36 Type-1 membership functions for the output variable

$$\mu\text{Normal\_Nocturnal\_Load}(x) = \begin{bmatrix} 0, x \leq 0 \\ \frac{x-10}{10}, 0 \leq x \leq 10.5 \\ \frac{22-x}{11.5}, 10.5 \leq x \leq 22 \\ 0, 22 \leq x \end{bmatrix}$$

$$\mu\text{Intermediate}(x) = \begin{bmatrix} 0, & x \leq 19 \\ \frac{x-30}{11}, & 19 \leq x \leq 30 \\ \frac{43-x}{13}, & 30 \leq x \leq 43 \\ 0, & 43 \leq x \end{bmatrix}$$

$$\mu\text{High}(x) = \begin{bmatrix} 0, & x \leq 39 \\ \frac{x-39}{31}, & 39 \leq x \leq 70 \\ \frac{100-x}{30}, & 70 \leq x \leq 100 \\ 0, & 100 \leq x \end{bmatrix}$$

Triangle interval type-2 membership functions with uncertain $a \in [a_1, a_2]$, $b \in [b_1, b_2]$ and $c \in [c_1, c_2]$.

$\tilde{\mu}(x) = \left[\underline{\mu}(x), \overline{\mu}(x) \right] = \text{itritype2}(x, [a_1, b_1, c_1, a_2, b_2, c_2])$, where $a_1 < a_2, b_1 < b_2, c_1 < c_2$

$$\mu_1(x) = \max\left(\min\left(\frac{x-a_1}{b_1-a_1}, \frac{c_1-x}{c_1-b_1} \right), 0 \right)$$

$$\mu_2(x) = \max\left(\min\left(\frac{x-a_2}{b_2-a_2}, \frac{c_2-x}{c_2-b_2} \right), 0 \right)$$

$$\overline{\mu}(x) = \begin{cases} max(\mu_1(x), \mu_2(x)) \; \forall x \notin (b1, b2) \\ 1 \hspace{3.5em} \forall x \in (b1, b2) \end{cases}$$

$$\underline{\mu}(x) = min(\mu_1(x), \mu_2(x)) \tag{4.15}$$

Input variables for triangular type-2 fuzzy system

1. **Diurnal_Load**. This variable has the following membership functions (MFS): Normal_Diurnal_Load, Intermediate_Diurnal_Load and High_Diurnal_Load. This MFS are listed in Table 4.37.

2. **Nocturnal_Load**. This variable has the following membership functions (MFs): Normal_Nocturnal_Load, Intermediate_Nocturnal_Load and High_Nocturnal_Load. This MFS are listed below in the Table 4.38.

Output variables for triangular type-2 fuzzy system

The FS has only one output called **LOAD_LEVEL**, which refers to the Blood pressure (BP) load level, the output gives a percentage, which refers to the following BP

Table 4.37 Type-2 membership functions for the first input variable

Normal_Diurnal_Load

$$\mu_1(x) = \max\left(\min\left(\frac{x-0}{7.01}, \frac{16.8-x}{9.79}\right), 0\right)$$

$$\mu_2(x) = \max\left(\min\left(\frac{x-5.42}{5.58}, \frac{20.7-x}{9.7}\right), 0\right)$$

Intermediate_Diurnal_Load

$$\mu_1(x) = \max\left(\min\left(\frac{x-18.6}{7.2}, \frac{36.8-x}{11}\right), 0\right)$$

$$\mu_2(x) = \max\left(\min\left(\frac{x-24.5}{5.5}, \frac{41.1-x}{11.1}\right), 0\right)$$

High_Diurnal_Load

$$\mu_1(x) = \max\left(\min\left(\frac{x-38.5}{25.6}, \frac{94.58-x}{30.48}\right), 0\right)$$

$$\mu_2(x) = \max\left(\min\left(\frac{x-44.8}{25.7}, \frac{100-x}{29.5}\right), 0\right)$$

Table 4.38 Type-1 membership functions for the second input variable

Normal_Nocturnal_Load

$$\mu_1(x) = \max\left(\min\left(\frac{x-0}{7.01}, \frac{14.42-x}{7.41}\right), 0\right)$$

$$\mu_2(x) = \max\left(\min\left(\frac{x-5.42}{5.58}, \frac{20.7-x}{9.7}\right), 0\right)$$

Intermediate_Nocturnal_Load

$$\mu_1(x) = \max\left(\min\left(\frac{x-18.6}{7.2}, \frac{33.73-x}{7.93}\right), 0\right)$$

$$\mu_2(x) = \max\left(\min\left(\frac{x-24.5}{5.5}, \frac{41.1-x}{11.1}\right), 0\right)$$

High_Nocturnal_Load

$$\mu_1(x) = \max\left(\min\left(\frac{x-38.9}{25.61}, \frac{90.3-x}{26.2}\right), 0\right)$$

$$\mu_2(x) = \max\left(\min\left(\frac{x-47.2}{23.3}, \frac{100-x}{29.5}\right), 0\right)$$

Load levels: Normal, Intermediate and High. The linguistic value of MFs is shown in Table 4.39.

Table 4.39 Type-1 membership functions for the output variable

Normal

$$\mu_1(x) = \max\left(\min\left(\frac{x-0}{7.01}, \frac{11.8-x}{4.79}\right), 0\right)$$

$$\mu_2(x) = \max\left(\min\left(\frac{x-6.74}{4.26}, \frac{20.9-x}{9.9}\right), 0\right)$$

Intermediate

$$\mu_1(x) = \max\left(\min\left(\frac{x-18.6}{7.2}, \frac{34.74-x}{8.94}\right), 0\right)$$

$$\mu_2(x) = \max\left(\min\left(\frac{x-26.9}{7.2}, \frac{41.1-x}{11.1}\right), 0\right)$$

High

$$\mu_1(x) = \max\left(\min\left(\frac{x-38.5}{25.6}, \frac{82.41-x}{18.31}\right), 0\right)$$

$$\mu_2(x) = \max\left(\min\left(\frac{x-52.2}{18.3}, \frac{100-x}{29.5}\right), 0\right)$$

4.4.6   Results

The experiments performed with the fuzzy system type-1 and type-2 for the classification of the blood pressure load have been good, currently, 100% accuracy rate was obtained when classifying the data of 30 patients as indicated in Table 4.40. In the column 3, we can see the pressure load levels based on the tables published by experts, and in column 4 and 5 we can see the result given by the type-1 and type-2 classifiers.

In the appendix, the tables corresponding to the pressure load module are shown, as well as the application thereof in the graphic interface.

4.5   Classification of Blood Pressure Level and Blood Pressure Load Using Bio-Inspired Algorithms: Genetic Algorithm (GA) and Chicken Swarm Optimization (CSO)

In previous works, different architectures of fuzzy classifiers were being carried out for the classification of blood pressure levels and blood pressure load, which were optimized by genetic algorithms and the best architecture found so far for 30, 60 and 80 patients. The best architecture for the blood pressure level classifier was: Triangular membership functions, 21 fuzzy rules, mamdani type and the best

Table 4.40 Results for type-1 fuzzy classifier

Patient	Diurnal_Load (%)	Nocturnal_Load (%)	Guideline (%)	Type-1 classifier (%)	Type-2 classifier (%)
1	2	55.3	High	High	High
2	8.3	47.8	High	High	High
3	55.2	41.8	High	High	High
4	8.0	45.0	Intermediate	Intermediate	Intermediate
5	77.0	82.9	High	High	High
6	0.0	2.3	Normal	Normal	Normal
7	76.0	33.3	High	High	High
8	0.0	5.0	Normal	Normal	Normal
9	53.0	56.1	High	High	High
10	32.0	23.0	Intermediate	Intermediate	Intermediate
11	46.9	93.9	High	High	High
12	54.2	90.0	High	High	High
13	62.4	23.0	High	High	High
14	3.3	33.8	Intermediate	Intermediate	Intermediate
15	28.6	57.3	High	High	High
16	36.8	21.2	Intermediate	Intermediate	Intermediate
17	7.9	5.3	Normal	Normal	Normal
18	5.3	14.7	Normal	Normal	Normal
19	46.8	18.8	High	High	High
20	17.7	31.8	Intermediate	Intermediate	Intermediate
21	9.9	27.6	Intermediate	Intermediate	Intermediate
22	36.3	16.7	Intermediate	Intermediate	Intermediate
23	9.3	0.0	Normal	Normal	Normal
24	7.3	31.3	Intermediate	Intermediate	Intermediate
25	55.5	0.0	High	High	High
26	71.6	47.1	High	High	High
27	23.6	33.3	Intermediate	Intermediate	Intermediate
28	5.8	15.3	Normal	Normal	Normal
29	61.0	63.0	High	High	High
30	0.0	12.0	Normal	Normal	Normal

architecture for the blood pressure load classifier was with triangular membership functions, 10 fuzzy rules and type Mamdani.

In this case, study, fuzzy systems were optimized for the classification of blood pressure levels and blood pressure load using genetic algorithms with different architectures and chicken swam optimization to compare the results once membership functions are optimized.

For this study, a database of 500 patients was used, which were taken from the Framingham database [19], with this data 30 fuzzy classifiers were analyzed for the classification of the blood pressure level, these generated by genetic algorithms as shows in Table 4.41 for type-1 and in Table 4.42 for the type-2 and another 30 fuzzy classifiers for blood pressure level classification using chicken swarm optimization, these experiments were performed using the same parameters specified in the Table 4.43 for type-1 and Table 4.44 for type-2. Subsequently, the same procedure was performed for the fuzzy blood pressure load classifier where the optimization of the triangular membership functions was created using genetic algorithms and subsequently using the chicken swarm optimization, these two configured with the same parameters to evaluate in order to perform a statistical study between the two and the comparative one is as fair as possible. Finally, Table 4.45 shows the results for type-1 and type-2 fuzzy systems using GA and CSO in a blood pressure level classifiers.

4.5.1   Statistical Test

To validate in the best way of the proposed method it is decided to use the statistical z-test, which is given by Eq. 4.13, and the parameters used for the test are shown in Table 4.46.

The following statistical test is to compare the results of type-1 and type-2 fuzzy classifiers optimized by Genetic algorithm as shown in Table 4.47.

The alternative hypothesis indicates that the Classifier Type-2 fuzzy systems optimized by GA with triangular membership function is greater than the classifier Type-1 fuzzy systems with triangular membership functions and the null hypothesis indicates otherwise, based on the information provided, the significance level is $\alpha = 0.05$, and the critical value for a right-tailed test is $z_c = 1.64$. Since it is observed that $z = 1.155 \leq z_c = 1.64$, it is then concluded that the null hypothesis is not rejected... Therefore, the claim of the alternative hypothesis is rejected, mentioning that the Type-2 classifier with triangular membership functions is greater than the Type-1 classifier with triangular membership functions. The "S" means that significant evidence was found and "N.S" refers to the fact that no significant evidence was found.

The following statistical test is to compare the results of type-1 and type-2 fuzzy classifiers optimized by Chicken swarm optimization as shown in Table 4.48.

The alternative hypothesis indicates that the Classifier Type-2 fuzzy systems optimized by CSO with triangular membership function is greater than the classifier Type-1 fuzzy systems with triangular membership functions and the null hypothesis indicates otherwise, based on the information provided, the significance level is $\alpha = 0.05$, and the critical value for a right-tailed test is $z_c = 1.64$. Since it is observed that $z = 3.261 > z_c = 1.64$, it is then concluded that the null hypothesis is rejected. Therefore, the claim of the alternative hypothesis is accepted, mentioning that the Type-2 classifier with triangular membership functions is greater than the Type-1 classifier with triangular membership functions. The "S" means that significant evidence was found and "N.S" refers to the fact that no significant evidence was found.

Table 4.41 Results for type-1 fuzzy classifier for the classification of blood pressure level using genetic algorithm

Genetic algorithm (GA) with type-1 fuzzy systems				
Parameters				
No.	Population	Generations	Dimensions	% of success of triangular MF
1	5	2000	72	88.2
2	10	1000	72	86.6
3	15	667	72	84.8
4	18	556	72	87.4
5	20	500	72	84.6
6	23	435	72	86.4
7	26	385	72	86
8	30	333	72	83
9	33	303	72	87.4
10	35	286	72	87.2
11	38	263	72	86.6
12	40	250	72	86.2
13	43	233	72	88.6
14	**47**	**213**	**72**	**89.6**
15	50	200	72	87.8
16	52	192	72	87.4
17	55	182	72	86
18	60	167	72	84.2
19	64	156	72	88.4
20	67	149	72	86.4
21	70	143	72	85.2
22	72	139	72	88.4
23	75	133	72	87.2
24	80	125	72	85.8
25	85	118	72	85
26	87	115	72	84.6
27	90	111	72	85.2
28	92	109	72	86.2
29	96	104	72	88.2
30	100	100	72	86
Average				86.42

Table 4.42 Results for type-2 fuzzy classifier for the classification of blood pressure level using genetic algorithm

Genetic algorithm (GA) with type-2 fuzzy systems

Parameters

No.	Population	Generations	Dimensions	% success triangular MF
1	5	2000	144	84.8
2	10	1000	144	85.6
3	15	667	144	87.6
4	18	556	144	87
5	20	500	144	89.6
6	23	435	144	84.2
7	26	385	144	88.4
8	30	333	144	87
9	33	303	144	86.8
10	35	286	144	87.2
11	38	263	144	83.4
12	40	250	144	86.8
13	43	233	144	87.4
14	47	213	144	85.8
15	50	200	144	83.2
16	52	192	144	83.8
17	55	182	144	86.2
18	60	167	144	86.4
19	64	156	144	85.6
20	67	149	144	87.4
21	70	143	144	87.2
22	72	139	144	89.4
23	75	133	144	88.4
24	**80**	**125**	**144**	**90**
25	85	118	144	88.8
26	87	115	144	85.8
27	90	111	144	88.8
28	92	109	144	87.8
29	96	104	144	87.8
30	100	100	144	87.2
Average				86.91

Table 4.43 Results for type-1 fuzzy classifier for the classification of blood pressure level using Chicken swarm optimization (CSO)

Chicken swam optimization (CSO) with type-1 fuzzy systems								
Parameters								
No.	Population	Generations	Dimensions	G	r%	h%	m%	% of success of triangular MF
1	5	2000	72	2	0.15	0.7	0.2	85.6
2	10	1000	72	5	0.15	0.6	0.2	86.4
3	15	667	72	18	0.15	0.8	0.3	86.4
4	18	556	72	4	0.15	0.6	0.3	85.6
5	20	500	72	8	0.15	0.5	0.1	89.4
6	23	435	72	5	0.15	0.7	0.6	86.4
7	26	385	72	7	0.15	0.8	0.3	89.4
8	30	333	72	13	0.15	0.8	0.2	87.8
9	33	303	72	9	0.15	0.5	0.4	87.2
10	35	286	72	13	0.15	0.7	0.5	87.4
11	38	263	72	2	0.15	0.6	0.1	90
12	40	250	72	11	0.15	0.8	0.5	86.4
13	43	233	72	3	0.15	0.8	0.2	85.8
14	47	213	72	14	0.15	0.5	0.3	86.8
15	50	200	72	10	0.15	0.8	0.4	87.6
16	52	192	72	13	0.15	0.5	0.5	86.6
17	55	182	72	3	0.15	0.6	0.4	86.4
18	60	167	72	2	0.15	0.7	0.3	87
19	64	156	72	1	0.15	0.8	0.2	87.8
20	67	149	72	5	0.15	0.7	0.1	87.6
21	70	143	72	6	0.15	0.6	0.3	85.6
22	72	139	72	3	0.15	0.8	0.4	87.2
23	75	133	72	9	0.15	0.7	0.2	86.2
24	80	125	72	15	0.15	0.6	0.1	88.8
25	85	118	72	19	0.15	0.5	0.3	90
26	87	115	72	7	0.15	0.6	0.2	86.4
27	90	111	72	9	0.15	0.7	0.1	89.2
28	92	109	72	9	0.15	0.8	0.3	90
29	**96**	**104**	**72**	**3**	**0.15**	**0.8**	**0.4**	**90.4**
30	100	100	72	12	0.15	0.5	0.3	89.2
Average								87.62

Table 4.44 Results for type-2 fuzzy classifier for the classification of blood pressure level using Chicken swarm optimization (CSO)

Chicken swarm optimization (CSO) with type-2 fuzzy systems

Parameters								
No.	Population	Generations	Dimensions	G	r%	h%	m%	% of success of triangular MF
1	5	2000	144	2	0.15	0.7	0.2	90
2	10	1000	144	5	0.15	0.6	0.2	90
3	15	667	144	18	0.15	0.8	0.3	91.2
4	18	556	144	4	0.15	0.6	0.3	91.2
5	20	500	144	8	0.15	0.5	0.1	88.2
6	23	435	144	5	0.15	0.7	0.6	92.8
7	26	385	144	7	0.15	0.8	0.3	90
8	30	333	144	13	0.15	0.8	0.2	88.2
9	33	303	144	9	0.15	0.5	0.4	87.4
10	35	286	144	13	0.15	0.7	0.5	87.2
11	38	263	144	2	0.15	0.6	0.1	86.6
12	40	250	144	11	0.15	0.8	0.5	88.2
13	43	233	144	3	0.15	0.8	0.2	91.2
14	47	213	144	14	0.15	0.5	0.3	91
15	50	200	144	10	0.15	0.8	0.4	87.8
16	52	192	144	13	0.15	0.5	0.5	91.2
17	55	182	144	3	0.15	0.6	0.4	88.2
18	60	167	144	2	0.15	0.7	0.3	90
19	64	156	144	1	0.15	0.8	0.2	91.2
20	67	149	144	5	0.15	0.7	0.1	93
21	70	143	144	6	0.15	0.6	0.3	90
22	72	139	144	3	0.15	0.8	0.4	90
23	75	133	144	9	0.15	0.7	0.2	87.2
24	80	125	144	15	0.15	0.6	0.1	91.2
25	85	118	144	19	0.15	0.5	0.3	91.2
26	87	115	144	7	0.15	0.6	0.2	91.2
27	**90**	**111**	**144**	**9**	**0.15**	**0.7**	**0.1**	**93.4**
28	92	109	144	9	0.15	0.8	0.3	93.2
29	96	104	144	3	0.15	0.8	0.4	91.8
30	100	100	144	12	0.15	0.5	0.3	88.6
Average								90.08

Table 4.45 Results for type-1 and type-2 fuzzy systems using GA and CSO in a blood pressure level classifiers

Results for type-1 and type-2 fuzzy systems using GA and CSO				
	Type-1 with GA	Type-2 with GA	Type-1 with CSO	Type-2 with CSO
Average:	86.42	86.91	87.62	90.08
Variance	2.27	3.11	2.22	1.96
Standard deviation	1.51	1.76	1.49	3.85

Table 4.46 Values for the statistical z-test

Parameter	Value
Level of confidence	95%
Alpha	0.05
H_a	$\mu 1 > \mu 2$ (Claim)
H_o	$\mu_1 \leq \mu_2$
Critical value	1.645

Table 4.47 Results of statistical test of type-2 fuzzy systems versus type-1 fuzzy systems with GA

Classifier	Zc	Z_Value	Evidence
Type-2 versus type-1 with GA	1.64	1.155	N.S

Table 4.48 Results of statistical test of type-2 fuzzy systems versus type-1 fuzzy systems with CSO

Classifier	Zc	Z_Value	Evidence
Type-2 versus type-1 with GA	1.64	3.261	S

The following statistical test is to compare the best architecture of GA with the best architecture of CSO; in this case, the best is Type-2 with triangular MFs using GA versus Type-2 with triangular using CSO as shown in the Table 4.49.

The alternative hypothesis indicates that the Classifier Type-2 fuzzy systems optimized by CSO with triangular membership function is greater than the classifier Type-2 fuzzy systems with triangular membership functions optimized by GA and the null hypothesis indicates otherwise, based on the information provided, the significance level is $\alpha = 0.05$, and the critical value for a right-tailed test is $z_c = 1.64$. Since it is observed that $z = 4.089 > z_C = 1.64$, it is then concluded that the null hypothesis is rejected. Therefore, the claim of the alternative hypothesis is accepted,

Table 4.49 Results of statistical test of type-2 fuzzy systems versus type-1 fuzzy systems with CSO

Classifier	Zc	Z_Value	Evidence
Type-2 with CSO versus type-2 with GA	1.64	4.089	S

mentioning that the classifier Type-2 fuzzy systems optimized by CSO with triangular membership functions is greater than the classifier Type-2 fuzzy systems with triangular membership functions optimized by GA. *The "S" means that significant evidence was found and "N.S" refers to the fact that no significant evidence was found.

References

1. Miramontes, I., Martínez, G., Melin, P., Prado-Arechiga, G. (2017). A hybrid intelligent system model for hypertension risk diagnosis. In *Fuzzy logic in intelligent system design, proceedings of the north american fuzzy information processing society annual conference, Cancun, Mexico, 16–18 October 2017*. Springer, Cham, Switzerland, 2017, pp. 202–213
2. Melin, P., Miramontes, I., & Prado-Arechiga, G. (2018). A hybrid model based on modular neural networks and fuzzy systems for classification of blood pressure and hypertension risk diagnosis. *Expert Systems with Applications, 107,* 146–164.
3. Guzman, J. C., Melin, P., & Prado-Arechiga, G. (2017). Design of an optimized fuzzy classifier for the diagnosis of blood pressure with a new computational method for expert rule optimization. *Algorithms, 10,* 79.
4. Słowiński, K. (1992). Rough classification of HSV patients. In R. Słowiński (Ed.), *Intelligent decision support. Theory and decision library (Series D: System Theory, Knowledge Engineering and Problem Solving)* (Vol. 11). Dordrecht, The Netherlands: Springer.
5. Yuksel, S., Dizman, T., Yildizdan, G., & Sert, U. (2013). Application of soft sets to diagnose the prostate cancer risk. *Journal of Inequalities Application, 2013,* 229.
6. Galilea, E.H., Santos-García, G., Suárez-Bárcena, I.F. (2007). Identification of Glaucoma stages with artificial neural networks using retinal nerve fibre layer analysis and visual field parameters. In E. Corchado, J.M. Corchado, A. Abraham (Eds.), *Innovations in hybrid intelligent systems. Advances in soft computing* (Vol. 44). Berlin/Heidelberg, Germany: Springer.
7. Alcantud, J.C.R., Santos-García, G., Hernández-Galilea, E. (2015). Glaucoma diagnosis: A soft set based decision making procedure. In J. Puerta (Ed.), *Advances in artificial intelligence, proceedings of the conference of the spanish association for artificial intelligence, Albacete, Spain, 9–12 November 2015*. Lecture Notes in Computer Science (Vol. 9422). Cham, Switzerland: Springer.
8. Alcantud, J.C., Biondo, A.E., Giarlotta, A. (2018). Fuzzy politics I: The genesis of parties. *Fuzzy Sets and Systems, 349,* 71–98.
9. Guzmán, J.C., Melin, P., Prado-Arechiga, G., & Miramontes, I. (2018). A comparative study between european guidelines and American guidelines using fuzzy systems for the classification of blood pressure. *Journal of Hypertension, 36.*
10. Zadeh, L. A. (1965). Fuzzy sets. *Information and Control, 8,* 338–353.
11. Yang, X. S., Karamanoglu, M., & He, X. (2014). Flower pollination algorithm: A novel approach for multiobjective optimization. *Engineering Optimization, 46,* 1222–1237.
12. Yu, J. J. Q., & Li, V. O. K. (2015). A social spider algorithm for global optimization. *Applied Software Computing, 30,* 614–627.
13. Mancia, G., Grassi, G., & Kjeldsen, S. E. (2008). *Manual of hypertension of the european society of hypertension*. Informa Healtcare: London, UK.
14. Wizner, B., Gryglewska, B., Gasowski, J., Kocemba, J., & Grodzicki, T. (2003). Normal blood pressure values as perceived by normotensive and hypertensive subjects. *Journal of Human Hypertension, 17,* 87–91.
15. Kaur, R., Kaur, A. (2014). Hypertension diagnosis using fuzzy expert system. In *International Journal of Engineering Research and Applications (IJERA) National Conference on Advances in Engineering and Technology, AET*, 29th March 2014.

16. Kaur, A., Bhardwaj, A., & Been, U.A.H. (2014). Genetic neuro fuzzy system for hypertension diagnosis. *Heart, 19*, p 25.
17. Poli, R., et al. (1991). A neural network expert system for diagnosing and treating hypertension. *Computer, 24*(3), 64–71.
18. Sikchi, S., & Ali, M. (2013). Design of fuzzy expert system for diagnosis of cardiac diseases. *International Journal of Medical Science and Public Healthcare, 2,* 56.
19. Framingham Heart Study (2019) [Online]. Available https://www.framinghamheartstudy.org/risk-functions/hypertension/index.php. Accessed 15 Jul 2019.

Chapter 5
Conclusions of the Neuro Fuzzy Hybrid Model

In this book, a new model has been designed using intelligent techniques, which implements human reasoning based on a set of decision rules based on an expert, with the aim of diagnosing different diseases, such as hypertension. This new model helps to have an accurate, fast diagnosis and above all to have a diagnosis based on the historical evidence of a patient in order for the expert to achieve a better decision. In this case the classification of blood pressure is studied following the definitions published by the European guides. The use of this model has been very efficient so far, so it has been shown that it takes less time and is more accurate to classify the blood pressure level. The implementation of systems with these characteristics in the medical area can greatly help the health system and personnel, especially in those countries that lack hypertension specialists, such as cardiologists, internists, among others. It is important to mention that this is only the beginning of the implementation of the neuro fuzzy hybrid model in medical diagnosis, in this case in cardiovascular diseases, particularly in hypertension. At the moment, in this research, ambulatory blood pressure monitoring has been conducted to carry out the study of blood pressure of patients for 24 h.

The model was proposed using a modular approach, which has been analyzed and experimented. First, we have the fuzzy classifier for the blood pressure level, which we can mention that it has given good results with the experiments performed so far. It should be mentioned that we need to have a more robust database to continue testing the fuzzy classifier, since this type of classifications has many possible cases and if the information analyzed does not fully cover all possible scenarios, we are omitting possible errors or possible successes in each case, and we suggest continuing experimentation and improve the database for future applications.

Subsequently, a classifier was made for the blood pressure load, which was optimized and experiments were carried out with the database that has been collected, but it is worth mentioning something similar to that of the blood pressure classifier,

© The Author(s), under exclusive license to Springer Nature Switzerland AG 2021   97
P. Melin et al., *Neuro Fuzzy Hybrid Models for Classification in Medical Diagnosis*,
SpringerBriefs in Computational Intelligence,
https://doi.org/10.1007/978-3-030-60481-3_5

which indicates the lack of data or the need to have a broader database that helps validate the classification in all possible real cases that may exist. As a conclusion regarding this module, we can mention that its classification is good since the ranges that are evaluated for the classification are wide and that allows not having to much uncertainty when making a decision.

In general, the performance of the neuro fuzzy hybrid model so far has been very good, but it is necessary to continue with more experiments and have more complete databases to be able to help the model to classify the information in the best possible way. As future work it would be to increase the number of patients in the current database. Apply the proposed model in other cardiovascular diseases to support the final diagnosis.

Finally, we hope that the work reported in this book and future work will serve as motivation to other researchers to work on artificial intelligent techniques for applications in medicine.

Appendix

In this section, the Graphical user Interface of the intelligent system for classification of Blood Pressure and Risk Diagnosis is shown for the reader.

Below is the graphical interface, which contains several modules, each module has a specific task for the diagnosis of blood pressure. Figure A.1 shows the general interface that shows the different modules for a complete diagnosis. Module one has the name of blood pressure level, and this module can give a manual or smart diagnosis. Figure A.2 shows the smart diagnosis for a person with normal blood pressure. Figure A.3 shows the smart diagnosis for a person with blood pressure Grade_1. Figure A.4 shows the smart diagnosis for a person with Grade_2 blood pressure and Fig. A.5 shows the smart diagnosis for a person with Grade_3 blood pressure. Finally, Fig. A.6 shows module two, which diagnoses the level of pressure

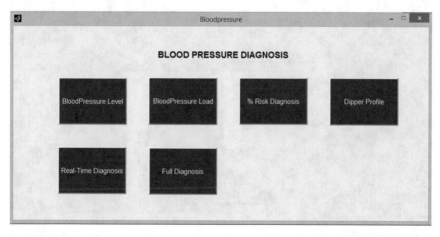

Fig. A.1 Graphical user interface of the intelligent system for classification of blood pressure and risk diagnosis

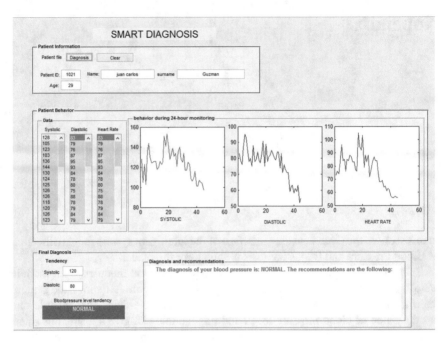

Fig. A.2 Blood pressure level with smart diagnosis for a person with normal blood pressure

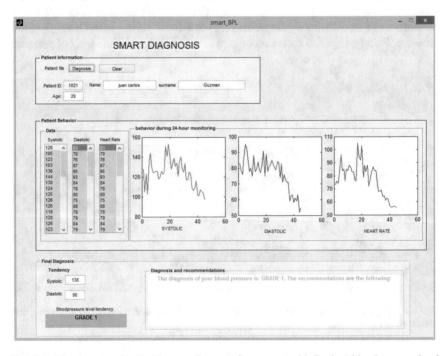

Fig. A.3 Blood pressure level with smart diagnosis for a person with Grade_1 blood pressure level

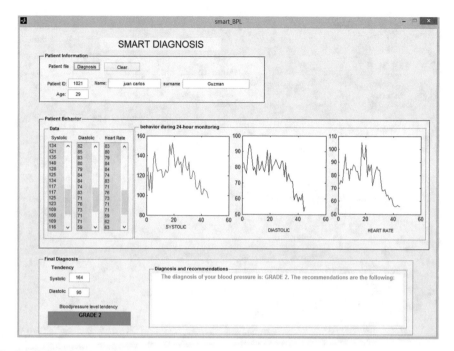

Fig. A.4 Blood pressure level with smart diagnosis for a person with Grade_2 blood pressure level

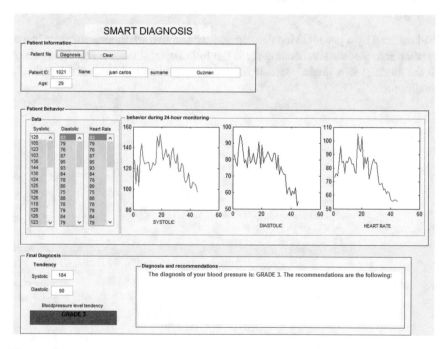

Fig. A.5 Blood pressure level with smart diagnosis for a person with Grade_3 blood pressure level

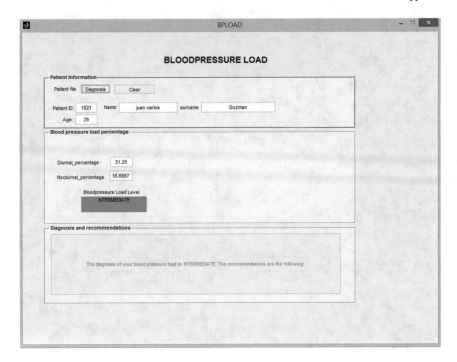

Fig. A.6   Blood pressure load module for a patient

load suffered by a patient. Modules, three, four and five are in tests for their incorporation into module six. Module six is the full diagnosis since it has the union of all the modules in a single interface to give the expert an overview of the patient's condition.

Index